ANCIENT

ANCIENT

Reviving the Woods That Made Britain

LUKE BARLEY

Profile Books

First published in Great Britain in 2026 by

Profile Books Ltd
29 Cloth Fair
London
EC1A 7JQ

www.profilebooks.com

Copyright © Luke Barley, 2026

1 3 5 7 9 10 8 6 4 2

Designed and typeset by Patty Rennie

Illustration copyright © Mike Biddulph, 2026

Printed and bound in Great Britain by
CPI Group (UK) Ltd, Croydon, CR0 4YY

The moral right of the author has been asserted.

Wendell Berry, excerpt from 'A Native Hill' from *Think Little: Essays*. Copyright © 2019 by Wendell Berry. Reprinted with the permission of Penguin Books Limited, and The Permissions Company, LLC on behalf of Counterpoint Press, counterpointpress.com.

Unless stated otherwise, all photographs are the author's own.

All rights reserved. Without limiting the rights under copyright reserved above, no part of this publication may be reproduced, stored or introduced into a retrieval system, or transmitted, in any form or by any means (electronic, mechanical, photocopying, recording or otherwise), without the prior written permission of both the copyright owner and the publisher of this book.

A CIP catalogue record for this book is available from the British Library.

Our product safety representative in the EU is BGC Sustainability & Compliance,
7 avenue du Général Leclerc, Paris, 75014, France
https://baldwinglobalconsulting.com

ISBN 978 1 80522 223 1
eISBN 978 1 80522 225 5

For Megan, Wren and William

Contents

	Prologue	ix
	Introduction	1
One	*The Pioneers*	7
Two	*Turn, Turn and Turn Again*	22
Three	*Being Cruel to Be Kind*	44
Four	*Common People*	69
Five	*A Glimpse of the Wildwood*	92
Six	*Woodbanks and Walking Trees*	106
Seven	*The New Dark Age*	133
Eight	*Darkness and Light*	159
Nine	*Inheritance*	192
Ten	*Treading Softly*	216
Eleven	*Wilding the Woods*	239
Twelve	*The Long Spring*	260
	Epilogue	285
	Further Reading	291
	Notes	292
	Acknowledgements	300
	Index	303

To walk in the woods, mindful only of the *physical* extent of it, is to go perhaps as owner, or as knower, confident of one's own history and of one's own importance. But to go there, mindful as well of its temporal extent, of the age of it, and of all that led up to the present life of it, and of all that may follow it, is to feel oneself a flea in the pelt of a great living thing, the discrepancy between its life and one's own so great that it cannot be imagined. One has come into the presence of mystery. After all the trouble one has taken to be a modern man, one has come back under the spell of a primitive awe, wordless and humble.

Wendell Berry, A Native Hill

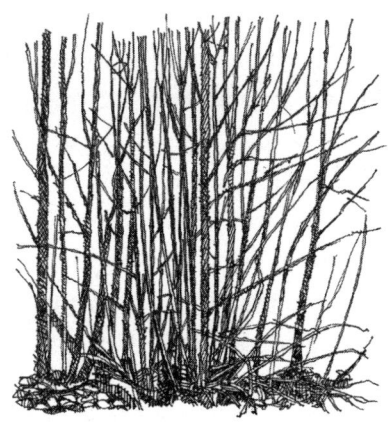

Prologue

RAIN TRICKLES COLD off my helmet and my shirt is so wet I might as well have jumped in the river that flows fast and brown far below. Even my legs are soaked, the thick protective fabric of my chainsaw trousers sodden, and as I kneel to make cuts in the base of another lanky ash the pale dust spewing from the saw clings to them. I step away as the falling tree cuts through the downpour and bounces and clatters to a stop.

With the long views down the dale lost in cloud, the detail at my feet springs into sharper focus. The battered orange casing of my chainsaw jars with the shifting, livid green moss, and both pop disconcertingly against the grey scree, smooth ash stems and sheeting rain. In rocky hollows, slimy ash leaflets rot ahead of their tough stalks, scattered like straw, while I pick out occasional leaves from a single, middle-aged oak. They vary in colour from tawny brown through copper to deep purple, and mark every stage of the oak leaves' multi-year deterioration back into the earth: this year's crop leathery and solid, older ones a lacework skeleton. A bright circle of last summer's foliage lingers on beneath a wych elm, glowing amber.

Kate, Andy and I huddle under a tarpaulin for lunch, stopping only for as long as it takes to wash down our sandwiches with tea from dented flasks. Our camp is in our favourite corner of the worksite, beneath a sprawling small-leaved lime that is almost a little wood in itself, a complex grove of interconnected life. At its heart is the tallest tree in this part of the wood, towering above our tarp from its perch on a low crag with pendulous limbs arcing away from the straight trunk to create a flowing, narrow crown. The lime is surrounded by a chaotic tangle formed by its own collapse and regeneration. One of the swooping branches has embedded itself in the loose earth and set root to spark a younger sister, while fallen stems – its former companions – remain connected at the base and have erupted along their length into rows of new trees, shooting for the light. A ragged stump has produced a dense mass of soft, orange-green twigs.

The tree – or grove, or thicket, whatever you want to call it – has a particular energy: it's not the solid, stoic presence of an old oak but something more supple and lithe, a victory of adaptation rather than resistance. The lime is irrepressible and every feature seems designed to survive, to persist and overcome. When it thrives, vigorous branches touch down and create a ring of offspring. Fall: the branches become trees. Fail, like that broken old stump, and dormant buds, visible in bumps and swellings up the stem, spring to life.

This particular lime could have been clinging on here since soon after the last Ice Age, roaming slowly around the steep slope with each new failure and reiteration. The bears and wolves whose bones have been found in the caves of this hollow landscape might have lain up in its shade and we speculate, on more leisurely lunch breaks, which of our ancestors might have sat beneath it and enjoyed its welcoming sense of enclosure as they took a rest from their own work. We won't touch it with our saws, but its presence roots our brief time here in a story of continual change from the deep past and into the distant future of this ancient wood. Lime has also, in some respects, brought me here: it has been a constant presence in my life as a ranger, a thread running between the most significant places I've come to know through the long accumulation of days like this, working hard to care for some of Britain's most special woods and landscapes.

But there's no point sitting about letting our soaked clothes grow colder and we get back into it with a shudder, pulling on chilled gloves. We fold the tarp without a word and tuck it behind a stump that has its own miniature grove of candle-snuff fungus emerging from a shroud of moss. The multiple generations of that single ageless lime rise and sprawl around us as we fuel the saws and head back to our work, sizing up the next trees for felling as we clamber across the scree.

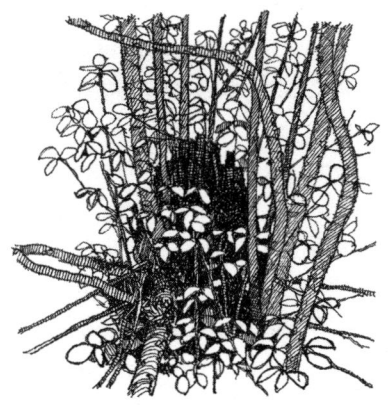

Introduction

PHYSICAL WORK GETS you under the skin of a place quickly. In woodland, work takes you off the path and into the undergrowth, up hillsides and down ravines and across hidden banks and ditches that quietly reveal lost histories. It forces you to become intimately familiar with the trees, soil, plants and fungi around which you're working, holding you in the same spot for hours and days – sometimes weeks – on end, revealing detail in the same way the grain emerges as you sand rough-sawn timber. This book tells my story of getting off the path and under the skin of some of Britain's most special ancient woods, in South London, the Lakes and the Peak District. As a ranger, I was responsible for every aspect of their care and spent my working life protecting them and improving their condition – both for the wildlife that makes a home among the trees, and for all the things woodland does for people. I've been involved with everything from the thinking work – the ecological surveys and management plans – to the doing: building fences, climbing trees and, more than anything, spending countless days felling.

Each ancient woodland that I've worked in has shown me more about these quintessential features of the British countryside which are defined by both their immense value for wildlife and their central role in human life since the Bronze Age. Ancient woods teem with wildlife from the soil to the canopy, forming our richest terrestrial habitats and providing a home for a quarter of Britain's priority species for conservation.[1] Some ancient woodlands are the last surviving remnants of the primeval 'wildwood', as Oliver Rackham called it, the landscape of Britain before people began to farm.

But these woods, now less than 3 per cent of our land area, are not simply the result of nature taking its own course. From the invention of stone tools until the middle of the twentieth century, the continuing existence and ecological health of these incredible places depended almost entirely on the influence of people. For millennia, the intensive management of woodland sustained the perfect conditions for all sorts of life to thrive, including human life: this was a foundational, interdependent relationship that has only recently been lost. The story of the woods in this book is not just one of trees, birds and butterflies, but a deeply human tale that traces the course of our history.

Until a couple of generations ago most of Britain's ancient woodland was bustling with human activity as our forebears 'coppiced' areas of trees, felling them for use and allowing the stumps to spring back to life with fresh new growth. Coppicing was an essential industry, providing almost everything people needed: seemingly insignificant scraps of wood were used as pegs and to hold thatch in place, bigger stems were carved into tools or bodged into furniture, and giant oaks grown over generations became ships and the roof beams of our grandest buildings. The soaring vaults of cathedrals typically needed more than 1,000 mature oaks for their frames, while even a modest medieval farmhouse would have required a couple of hundred trees of various sizes, as well as hundreds of hazels for its wattle and daub walls – and that's before we consider the continual need to fuel its hearths and ovens. The sustainable production of firewood underpinned all human life, while charcoal – made by baking the moisture out of coppiced wood – was essential to the development of metals as it provided fierce, steady heat for smelting.

Vast other areas of the British countryside – now neatly compartmentalised into agricultural fields, forestry plantations and remnant woodland – were managed as another type of ancient woodland where grazing animals continued to roam beneath 'pollards', trees cut above the reach of hungry cattle and that similarly grew back to produce a renewable crop of wood. 'Wood pasture' is home to some of our oldest trees and the extraordinary range of life they support, as well as forming an even less well-understood – but equally important – theme in the fundamental story of people and woodland.

Over the last two hundred years or so, the Industrial Revolution and the rise of fossil fuels have driven most of our ancient woodland to destruction and abandonment. The increasing availability of coal and oil, along with the development of plastic, modern bricks and steel, made most wood products redundant. It became more efficient to produce those that remained, like the roof beams of houses, from industrially managed forestry plantations. Between the end of the Second World War and the 1980s nearly half of Britain's ancient woodland was cleared or replanted with commercial crops of trees, while the area of woodland under coppice management plummeted from over 230,000 hectares in 1905 – when it was already dwindling – to less than 25,000 hectares in the 1990s, leading to precipitous declines in the wildlife that had come to rely on the light, warmth and varied structure of coppiced woods.[2]

Most ancient woodland is now protected from clearance thanks to the efforts of pioneering conservationists in the 1970s and 80s, but some woods continue to be at risk from built development. More, however, are at risk from other pressing threats like tree disease, climate change, a booming and hungry population of deer and, most significantly – and perhaps counter-intuitively – neglect. Many ancient woods, abandoned once their produce was no longer needed, require careful human management to revitalise their waning ecosystems and save their remnant specialist species: the Forestry Commission found that only 58 per cent of Britain's total woodland area is under active management, with ancient woods almost certainly faring much worse. Less than 10 per cent of native woodland is in good ecological condition.[3]

But all is not lost. This book also tells the story of the friends and colleagues I've worked with, modern woodland managers who are revitalising the long traditions we can read in the current shape of the woods, even as we write the next chapter; not, perhaps, with our axes, as pioneering conservationist Aldo Leopold proposed in the 1940s, but certainly with our chainsaws and planting spades. Like our forebears with their exhaustive use of coppice products and timber, we need twenty-first-century woodlanders to get under the skin of their places and to know and love them in a practical and unsentimental way, a physical engagement of bramble scratches, sore shoulders and the crash of falling trees. We'll also need bold thinking that reimagines the types of things we take from our woods and the ways we get them, broadening the scope of woodland management to encompass the reintroduction of missing keystone species and making some of our woods wilder, even as we rediscover our deep human connection with them.

After introducing the basics of ancient woodland, this book charts our heritage as a woodland species through the timeless traditions of coppicing and wood pasture. I then consider some of the threats facing our ancient woods, before articulating why a renewed engagement with ancient woodland is so important, moving roughly from the distant past to a potential future. With the climate changing and nature in crisis, the story of our

INTRODUCTION

ancient woods holds the key to a better relationship with the natural world. A resurgence of the wood culture that has been so integral to our history will play a crucial role in the survival of both the human species and our shared ecosystem.

I've been fortunate to work in dozens of special woods as a ranger and, for the last few years, as an adviser for the National Trust, where I've been involved with the care of woods across England, Wales and Northern Ireland. There are three in particular, though, that had the biggest impact in developing my connection to trees through the grimy intimacy of practical work, and which proved invaluable to my understanding of the ecology and human history of ancient woodland. Between them, they illustrate the wider story of the decline in traditional management and the potential benefits of its revival.

Spring Park in South London is a suburban coppicewood with a readily legible history of everyday use by our ancestors and a lively modern tradition of conservation; it's where I first learned the ropes of woodland work. Nearby Ashtead Common is a relic wood pasture where I climbed ancient oak pollards to carry out essential conservation pruning and witnessed first-hand the restoration of its grazed landscape; a deeply human place that simultaneously provides a glimpse of the pre-agricultural wildwood. And Dovedale Wood is a lost world nestled among the limestone crags of the White Peak in Derbyshire and proved my biggest challenge as a ranger: it's one of the richest woods in the country for wildflowers but faces an existential threat from ash dieback, a virulent fungal disease.

Spring Park and Dovedale are connected by a strong thread that ran through my practical career: they're both home to small-leaved lime, an uncommon tree species whose particular mechanisms for survival make it one of the most powerful indicators that a wood is ancient and entwine it with the length of human history. Now little known, lime once dominated the treescape of parts of Britain, and the places we find it today give us a rare glimpse of the wild on this tame island. We'll also visit other woods where this charismatic species clings to life, like Dodgson Wood in the Lake District, to consider how the saga of lime connects us with both the deep past and a possible future ecology, both within and outside the human story.

This is my story of Britain's ancient woodland. Like all writing on the woods, this book – and my understanding – owes a huge debt to a few greats of woodland ecology and history, most notably Oliver Rackham, George Peterken, Charles Watkins, Peter Marren, Keith Kirby and Donald Pigott. It's informed as well by other writers who bring a more diverse perspective to our relationship with nature; and, in addition to these literary figures, I've also been incredibly fortunate to learn from the many colleagues who've been generous in sharing their expertise throughout my career.

George Peterken told me that 'the story of ancient woodland is always many-stranded: there isn't a strong plotline'. In attempting to make this complex history understandable, I've inevitably oversimplified some nuanced points and glossed over some of the myriad parallel threads in the story of the British landscape. But, while I'm indebted to those experts, this book is unashamedly personal: it shares my experience of woodland work, reviews the state of ancient woodland today as I see it, and considers how a renewed engagement with this special habitat could contribute to a better future for both human society and the wildlife of the woods. This personal perspective extends to the fact that nothing I write should be taken as a National Trust position, but is simply my own view on things.

CHAPTER ONE

The Pioneers

Britain's ancient woodland and how it was saved

BEFORE I KNEW anything about their history or wildlife, it was stories that drew me to ancient woodland; not that I knew what ancient woodland was as a ten-year-old. As a child I followed my imagination to the woods in search of Robin Hood and King Arthur, the Swallows and Amazons, Tolkien's Ents and – perhaps even more often – the Ewoks of *Star Wars*. Even then, I knew that some woods were somehow more special than others: more atmospheric, with gnarlier trees, banks of bright wildflowers in spring, and mysterious lumps and depressions. I've heard it said more recently that, if you can imagine a fairy popping out from behind a tree, you're probably in an ancient wood. I was lucky to know a few woods like this around Marple, the town where I grew up, right where south Manchester meets the edge of the Peak District. But the one that came to mean the most to me was in the narrow valley behind the house of my friends Charles and Eddie, where the three of us, along with my brother Jake, spent endless days exploring and engaged in violent reenactments of our favourite books and films.

It was all too easy to believe that mysterious beings might pop out here, although for us they were most likely to be Kevin Costner's outlaws from *Robin Hood: Prince of Thieves*, or the Brownies of *Willow*. Being in what we came to call Eddie's Wood just felt good, whether I was exploring with the gang or sitting alone in a big oak, and it's a sensation I can still recapture by stepping into an old woodland.

The wood belonged no more to the brothers than it did to me, but we would have thought absurd the idea that someone owned it, lying as it did directly behind their garden without even a fence to separate them. If you've ever played or walked in a clough like this – or, elsewhere in the country, in a dell, dingle, ghyll, dene, combe or hollow – you'll know the type of wood I mean. It was long and thin, running pretty much unbroken for a couple of miles along a narrow, steep-sided valley, the wood seemingly bubbling over into the gardens it passed at its lower end and the sheep fields further up. The slopes shimmered with bluebells in April and there were other flowers, white and yellow, that came earlier but that I wasn't able to name as a child; I found them even more magical as they nodded in the low late winter sun, a secret sign of the imminent spring.

Tall, spreading oaks – the only real tree I could recognise – formed the backbone of a dense and shady canopy over gloomy patches of holly and other, nameless trees. The paths, such as they were, were beaten only by the feet of the occasional dog-walker or gangs of kids like us, and they followed the lazy little stream and the top edges of the valley along sagging barbed wire and past piles of grass cuttings dumped over collapsing fence panels. Downstream there was, then, a derelict mill, the tower of which was accessible by a death-defying leap from a stone retaining wall to a window ledge. The wood was in the process of reclaiming the dry millpond with thick, flowing roots that consumed its crumbling walls and leats; it was our Angkor Wat, imbued with a sense of deep time.

I loved Eddie's Wood. Along with other woods I knew, it was not only a place of adventure but a refuge from the confusing business of growing up, a role that woods continued to play for me throughout my teenage years and arguably still do today. I assumed then that places like this were permanent features of the landscape, sacrosanct even. It would have seemed incomprehensible to the ten-year-old me that, as we conducted our expeditions in the clough, Britain's foremost woodland experts were exploring similar woods with a much higher purpose in mind: ensuring their very survival.

In the 1970s the old woods like Eddie's – the native woods of oak, beech and ash – were under grave threat. Modern forestry was in the ascendant and encouraged by hefty tax breaks; since the end of the Second World War old woods had been planted over with monocultures of conifer or simply demolished to make way for agriculture, which was in its post-war drive to maximise production with the memory of rationing still strong. In an age of rapid technological progress, the trend was towards commercial or 'scientific' forestry over the woodmanship of earlier generations. This was a quest for increased efficiency, standardisation and profit over the nuanced and labour-intensive nurture of Britain's native woods, which were increasingly dismissed as useless scrub.

Oliver Rackham, the late, pre-eminent historian of Britain's landscape and one of the authorities on woodland, called this period the 'Locust Years' for ancient woods, reflecting the fact that it was a time of comprehensive destruction. Their traditional management – and with it their direct utility for society – had declined steeply with the rise of coal and oil, the availability of cheaper steel and other materials, and, later on, the industrial output of timber from plantations. While the fairly new nature conservation movement recognised their importance for wildlife, it struggled to protect the old woods as it grasped to articulate and define exactly what made them so significant. And although the public increasingly enjoyed visiting bluebell woods on the edges of towns and cities (a pursuit that had been advertised on railway posters since the Victorian era), there was no name nor criteria to define these special woods, and little sense of their value in public or academic discourse. Much of the clearance and conversion was happening in rural areas, where it went unquestioned if it was observed at all, in a time before every family had a car and with much of the countryside still strictly closed to the public.

Rackham and George Peterken, then of the Nature Conservancy Council, had begun to realise through their fieldwork during the 1960s that the woods with the highest importance for nature conservation – that is, where the most species, and the rarest species, lived – were those with the longest continuity of native woodland cover. Rackham, an extraordinary scholar and radically original thinker, published his first book, *Trees and Woodland in the British Landscape*, in 1976, describing the cultural and ecological significance of the old woods and revolutionising the way that many people involved with woodland and the countryside thought about them.

Despite his brilliance, though – or perhaps as an aspect of it – Rackham was anything but tactful, and his mischievous criticism failed to affect the attitudes of the foresters who had the power to influence the protection and management of nature-rich old woods. Peterken, on the other hand, not only recognised the threat but was able to do something about it, working as he did for the government's nature service and also possessing the diplomatic nous and political savvy to effect change. He and Rackham realised that, in order to protect these special places, they needed to be able to define and identify them, and so they agreed the criteria based on their extensive site visits and historical research. At its simplest, they decided, a wood is 'ancient' if it was already in existence in 1600:* at this date woodland is often shown on the earliest accurate maps and, if a place was wooded 400 years ago, it was less likely to be the product of tree planting – a later development – and more likely to have remained wooded, in some form, since trees recolonised Britain after the last Ice Age. Woods known to be in existence in 1600 could be direct descendants of the primeval wildwood, the more natural landscape before humans began to farm; even where they've grown on land that was previously more open, they've retained or developed incredibly rich ecosystems, which have benefitted from centuries of tree cover and the fact that the vibrant woodland soil hasn't been ploughed or received any other agricultural 'improvement' for hundreds of years.

Rackham's genius was in his study of both the ecology of woods and their human histories, capturing the imagination of his readers with his observation that not only are ancient woods vital for nature, they are also, in most places, a human artefact that pre-dates the parish church. Almost everyone I've spoken to while thinking about this book – including some of those pioneers like George Peterken and his old colleague Charles Watkins, another landscape historian – has emphasised that it's the combination of the habitat's rich ecology with its significance for people that draws them inexorably back to ancient woodland (and another, younger woodlander, Jamie Simpson, told me years ago that becoming interested in woodland is like joining the Mafia: once you're in, there's no way out). These are deeply human places, so it's no wonder that the young me was drawn by the stories held by Eddie's Wood and others like it. The concept of 'ancient woodland' as a formal label, though, dates from just a few years before I began exploring them.

*

---

* 1750 is used in Scotland, which is the date of the earliest accurate maps there.

Following Rackham and Peterken's criteria, woodland that has developed on open land since 1600 is not considered ancient and, if it developed naturally, is – in the simplest terms – called 'secondary woodland' (this has normally happened on former farmland, but can also be where land was cleared for other purposes, like mining or quarrying). Secondary woodland is often still very valuable for wildlife, not to mention for carbon storage and other ecosystem services, so shouldn't be maligned: but it doesn't have the same profound importance as ancient woodland. The continuity of the primeval woodland soils has been disrupted – typically by ploughing and fertilisation – and many woodland species can't readily recolonise these 'new' woods, particularly the fungi, wildflowers and invertebrates that don't have the mobility of birds and mammals.

There are also plenty of places that have been wooded since the seventeenth century but which were more open before that. We know now, for instance, that, prior to the Norman era and their formalisation into intensively managed coppicewoods, many ancient woods were in fact wooded commons and deer parks, while recent satellite imaging has even shown ancient field patterns under lots of old woods. A woodland can therefore – confusingly – be both ancient and secondary, and indeed this might be the case for the majority of ancient woods that developed from 'wood pasture', where livestock or large wild herbivores grazed among the trees for centuries. Regardless of their history of tree cover, though, which has waxed and waned in many ancient landscapes, the most important sites have simply not been ploughed or received other intensive agricultural interventions for hundreds of years, if ever.

The majority of the existing woodland in Britain today has not developed naturally but is 'plantation', which means the trees have been deliberately planted. Plantations exist on a spectrum from those that are barely discernible from natural woodland, as in the case of sensitively managed stands of native broadleaved trees, through to unnatural impositions in the landscape like the uniform monocultures of non-native conifer that form the least natural – but most commercially important – extreme of the modern forestry industry. Many ancient woods received some planting in the Victorian era, when the value of coppice began to wane and thoughts turned to timber production; where these plantations are of native species it can be difficult to discern them from natural woodland, and the ecological value of the wood may not be affected. Very many other ancient woods, though, were densely planted in the decades after the Second World War with conifer, which shaded out light-demanding woodland wildflowers and acidified the soil through needle-fall, suffocating their ecosystems.

It's possible, now, to find old maps online that can help determine the age of a wood. I looked for Eddie's Wood and, while I couldn't find maps dating back to 1600, it's present and correct on the first Ordnance Survey map of the area from the mid-nineteenth century, with neatly stamped symbols of broadleaved trees filling much of the length of the sinuous valley. It's labelled there 'Slack Wood', a name lost to time. A 'slack', though, is a depression in the landscape, so its old name is indicative of one aspect that makes this wood typical: much remaining ancient woodland is confined to the steepest slopes and tightest valleys, places that were, as Oliver Rackham pointed out, not necessarily good for trees but no good for anything else. Woodland survived in slivers and patches on ground that was too steep to plough, or on the least desirable land for grazing.[1]

But that isn't to say it was useless. Woods were essential to people and the fact that any one wood still exists today demonstrates that fact: there are plenty of other cloughs like this nearby that were cleared of trees and grazed. The survival of ancient woodland to the twentieth century is nearly always related to historic local needs for specific wood products. The old maps I found online show that the derelict mill in Eddie's Wood was a cotton mill, typical of the wooded valleys around Manchester where the constant rainfall and humid microclimate provided the ideal conditions for processing and weaving textiles. Built in the second half of the nineteenth century, this mill would almost certainly have used coal to power its furnace but, along with its hundreds of counterparts, would still have used a wide range of other essentials made from native wood even during the industrial heyday of Manchester cotton: bobbins to fit out the looms, tool handles, the components for carts and narrowboats, and clogs and all sorts of other domestic items for the workers.

Earlier enterprises were even more reliant on the sustainable supply of local firewood and charcoal, and the survival of many old woods, including our biggest remaining concentrations like those in the High Weald and the Forest of Dean, is easily attributed to major historic industries like iron smelting and tool- or furniture-making, or to their proximity to busy communities like abbeys and priories. In other places, like here in Mellor, it's harder to be certain exactly what the wood was managed for – but we can be sure it was essential for something.

Due to the constant historic use of the woods by people, the new concept of ancient woodland was properly labelled by the Nature Conservancy Council 'Ancient *Semi-Natural* Woodland' (ASNW), emphasising the fact that extremely little ancient woodland in the UK is genuinely entirely natural; it has nearly all been managed intensively, albeit sustainably, by people, and its extent, structure, and the range of tree species within it have been heavily modified to provide the things we needed. Even those places where people didn't need – or weren't able – to cut down trees, like on the most inaccessible slopes of the uplands, have been changed dramatically by the way we've grazed livestock. As we shall come to see, the involvement of people became an essential part of healthy woodland ecosystems in Britain, supporting animals and plants that evolved to thrive in the dynamism and variety of the pre-agricultural wildwood as generations of woodlanders unwittingly replicated similar conditions.

THE PIONEERS

*

That I can't readily find any maps of Mellor earlier than 1842 demonstrates the main shortcoming in relying solely on historical documents to define ancient woodland: the maps simply don't always exist, and, even where they do, the earliest ones can be hard to get hold of in libraries and physical archives. During his fieldwork in Lincolnshire in the 1970s, George Peterken realised that there was a suite of plants that were largely restricted to the area's ancient woods, and he called them 'ancient woodland indicators'. Over the subsequent couple of decades, as the profile of ancient woodland grew, most counties developed their own list of indicators, recognising the fact that flora varies with climate, geology and other factors – although there are perhaps a couple of dozen species that might be found in ancient woodland almost anywhere, like wood anemone and herb paris. As the name implies, the presence of these plants in a wood suggests that it is likely to be ancient, typically due to the fact that they can't easily recolonise younger woodland because their seeds only disperse a very short distance, limiting their ability to spread. Other indicator species don't reproduce by seed at all but rely on 'vegetative reproduction', where new plants grow directly from the roots or layering branches of the parent.

The most iconic and widely known species on the list, and the only one that I would have recognised in Eddie's Wood as a child, is the English bluebell. Bluebells are, though, one of the weaker indicators as they *are* able to spread into new woods more rapidly than some of the others (although still slowly in human terms); their seeds can spring a few metres when the dry flowerhead is brushed by a dog or deer. Ecologists typically expect to count at least ten of these weaker indicators to build an impression that a wood might be ancient.

At the other end of the spectrum, some species are very rarely found outside ancient woods and are therefore strong indicators of a woodland's age. The seeds of the wood anemone, for example – the white flowers that miraculously appeared to herald the end of winter in Eddie's Wood – typically land less than twenty centimetres from the flower, while for the last few thousand years small-leaved lime trees have only been able to persist by reinventing themselves through vegetative reproduction. The warming climate means that we do now find lime seedlings in British woods (and their bracts helicopter away on the wind like sycamore seeds, potentially colonising new places), but woodland conservationists can read the complex form of older lime trees – like the connected thicket we worked around in the prologue – to interpret their age and significance. Lots of other species are restricted to ancient woodland, from wasps and beetles to lichens and fungi, but plants are most helpful; firstly because 'they don't fly away', as woodland ecologist Keith Kirby writes, but also because they are relatively easy to identify in comparison to the more obscure groups that require greater expertise to understand properly.[2]

That it's possible to define this list of indicator species by their poor ability to recolonise newly planted woods also goes some way towards illustrating why ancient woods are so important in terms of the wildlife they support. Not only do they form our richest terrestrial habitat, but they are also impossible to recreate, on human timescales at least, and therefore irreplaceable. Even planting or translocating many of the flowers from the indicator list has proven difficult or impossible, and we also know that much of the value of ancient woodland lies beneath the plants, in the soil that's been maturing undisturbed for at least half a millennium and forms an incomprehensibly rich ecosystem; worms and beetles are the giants among a thriving fauna of microscopic organisms. Billions of tiny soil invertebrates, nematodes and protozoa live among the mind-bogglingly sophisticated network of fungi that ties the life of a wood's vegetation together. The life of the soil can be destroyed in a day by ploughing but the loss of trees doesn't necessarily affect it, whether by natural failure or human felling. Not only is it normal for trees to come and go while the soil beneath continues to develop and support a huge range of species, but the soil may remain intact in some of the woods that have been felled and replanted with conifers since the Second World War, making their restoration both possible and vitally important.

Having established a name and definition for ancient woodland, the pioneers at the Nature Conservancy Council launched a project to identify Britain's old woods and created, by the end of the 1980s – when I was exploring Eddie's Wood – a list of all the ancient woodland in the UK. This was an extraordinary undertaking that involved the study of old maps and documents as well as field surveys in search of those ancient woodland indicator species and other clues to the antiquity of woods, like charcoal hearths and woodbanks. In my mind the project holds the romance of Indiana Jones, with its combination of days spent in dusty libraries followed by the exploration of some of Britain's oldest and most mysterious places. The resulting Ancient Woodland Inventory, or AWI, listed around 35,000 individual woods across England, Wales and Scotland, covering around 2 per cent of our land area at just under 500,000 hectares (Northern Ireland wasn't included in the original survey).[3]

The surveyors estimated that 50,000 hectares had been lost to agriculture and development even since the Second World War, while, of what remained, a third had been converted to plantation forestry. Half had been abandoned and only a tenth was still under sensitive traditional management like coppicing. Two-thirds of the woods listed were considered 'small', which in woodland terms means less than five hectares in size (about seven football pitches), and increases their susceptibility to various threats.

These sobering statistics – as well as tireless advocacy work – led to amazing progress in the protection of old woods; it's no exaggeration to say that the work of Rackham, Peterken and their colleagues laid foundations that ultimately saved Britain's ancient woodland, radically slowing the pace of clearance and all but ending the practice of over-planting old woods with non-native conifers. Major changes in Forestry Commission policy and planning law followed that gave ancient woodland further protection, and even saw the commission begin to shift away from its original remit of just growing timber and towards providing society with a wider range of the benefits of woodland, like nature and public access. Derek Niemann, in his history of their endeavour, wrote that, 'Everyone who was there at the time agrees that these two visionaries, different and complementary, working separately and in partnership, achieved a seismic change by combining dogged determination with creative brilliance.'[4] George Peterken told me simply that the AWI and his influence on the Forestry Commission's change in direction were collectively 'the best thing I ever did in conservation'.

Protecting Ancient Semi-Natural Woodland is now a core principle of the conservation and forestry sectors, and the classification of a woodland as ASNW influences everything from applications for felling licences to the availability of grant funding. Although some landowners originally resisted the designation of their woods as ancient, fearing it would limit their freedom to clear them or to plant more lucrative tree species, Charles Watkins reflected recently that 'the development of ancient woodland as a concept worked remarkably well in redefining people's relationship with these old woods, and in protecting them from development; people are proud of their ancient woodland now'.

This isn't to say that there's no threat of clearance: the first phase of the HS2 trainline, for instance, led to the destruction of parts of more than fifty ancient woods, and the Woodland Trust's campaigns team estimate that between 2014 and 2024 more than 1,000 woods were threatened by development.[5] In comparison to the rate of destruction between the 1950s and 1970s, though, ancient woods have become, for the most part, the sacrosanct places I assumed them to be as a child.* But as we shall see, there are other threats to the health of their ecosystems beyond simply their eradication.

Thirty years on, responsibility for the Ancient Woodland Inventory is devolved to the nature conservation arms of the governments in England, Wales and Scotland. Each has modified their criteria and approach in the intervening time, which can make understanding the picture across the United Kingdom difficult, but the Woodland Trust attempted an overview in their 2021 *State of the UK's Woods and Trees* report. They estimate that there is today just over 600,000 hectares of ancient woodland in the UK, the increase from the original inventory being a result mainly of continued survey and improved data collection. This forms 19 per cent of the UK's total woodland cover, just 2.5 per cent of our land area. The rest of the UK's woodland is conifer plantation (51 per cent), with the remaining 30 per cent being secondary woodland and broadleaved plantation. Almost 40 per cent of ancient woodland was felled and replanted in the Locust Years, much of it with conifer; these areas are now classified as 'Plantations on Ancient Woodland Sites', or 'PAWS', and their restoration is a conservation priority.[6]

Between 2019 and 2025 the records for every English county were refreshed and renewed, with specialist consultants working with Natural England, the county records offices and other partners to review first-edition maps and other documents, as well as visiting around 10 per cent of the woods they identified to ground-truth their research. The scale of the original undertaking is demonstrated by the fact that updating each county took two or three years and the project cost £2.5 million. The

---

* As I put the finishing touches to *Ancient* in May 2025, the government is considering tearing up essential legislative protection for nature to enable easier built development. We may yet have another fight on our hands to save some of Britain's most important woods.

review has also brought thousands of the smallest woods into scope – the original inventory only recorded woods bigger than 2 hectares – as well as more historic wood pasture, a much more challenging type of ancient woodland to identify on old maps.

Anyone can access the inventory for England on the government's Magic Map website or through the Forestry Commission's online mapping service. (Scotland's data is similarly held on the NatureScot spatial data hub, while Natural Resources Wales hosts theirs on DataMap Wales.) Zooming in to Eddie's Wood on Magic, I see that much of it is shaded with the vertical green lines marked in the legend as Ancient Semi-Natural Woodland. Knowing what I know now about its location, its history and its banks of bluebells and wood anemones, this is no surprise. I also see that Marple, my hometown, is surrounded by a disproportionate amount of ancient woodland, including many of my childhood hangouts: Brabyns Park, where I jumped my BMX; the wood behind Matt and Doug's houses, where I'd nag to turn off the computer and explore; and, most extensive of them all, Etherow Country Park, above which my friend Olly's family lived on an old farm and where as teenagers we continued to climb increasingly nerve-racking trees and build elaborate dens. I loved all of those woods for reasons it's hard to articulate now without applying my adult knowledge, but I think the ten-year-old me would have agreed that they spoke to me with their old, Tolkien-esque oaks and the banks of ferns that conjured Endor and the Ewoks, with enchanted, changing patterns of wildflowers, and with their feeling of great age. They were places of fun and adventure with friends, but also of solace and refuge, and they always felt like home.

Perhaps, then, it was inevitable that I'd end up a professional woodland nerd. I certainly liked the idea of it as I toiled to build dens and treehouses, or when I saw rangers in their Land Rovers while on walks in the Peak or on holiday in North Wales. But however much I daydreamed of working in the woods, I had no idea what it actually meant, didn't know anyone who did it and didn't have the first clue how to get started. I followed a circuitous path through my teenage years from Eddie's Wood to finding a job as a ranger, and it would take a trip to the redwoods of California to eventually provide the motivation I needed to come back and immerse myself in Britain's ancient woodland – and experiences that would have amazed me as a child.

When I made the leap into conservation I assumed that I would somehow be involved with 'saving' woodland, like my 80s generation were exhorted to save the tropical rainforests. Having learned that the efforts of Oliver Rackham, George Peterken and co. had already halted the loss of Britain's ancient woodland – for the most part – but that its wildlife remained in precipitous decline, I ended up instead involved with aspects of its care that were more nuanced and unexpected, although no less vital. A new generation of woodlanders now builds on the incredible legacy left by the pioneers of the 1970s and 80s by rejuvenating the techniques that humans used to care for the woods from the Bronze Age to the Second World War. We want to ensure that our ancient woods aren't simply preserved, vital as that is, but that they return to being the dynamic, thriving ecosystems – and vibrant human places – that they were throughout history.

CHAPTER TWO

Turn, Turn and Turn Again

Six millennia of coppicing

EACH PULL OF my pruning saw produces a little firework of bright dust as I lean into the hazel bush, shoulder set against a thicker pole, and reach around to work where the twiggy tops of the stems I'm cutting will fall neatly away from the crown. I whip off the slimmest rods first, clearing the way: nut brown, pencil-straight and the width of my thumb at their base, they each take no more than a couple of strokes. The dozen other narrow poles are more like the size of my wrist and their bark at this age is paler, the colour of last year's oak leaves but with the sheen of burnished bronze. I nick the front face of each stem with the saw, then, as I work in from the back, hold it with my left hand to stop it splitting out early and – hopefully – guide its gentle fall on to the pile.

Some of them won't fall even once cut, their branches grown in a tangle around those of their neighbours. I put my saw down to maul them out, pulling on their base with both hands and enjoying the sensation of the smooth bark with its roughening pattern of lenticels, the raised pores that allow the wood of the tree to breathe. The coarseness is matched by a background of white noise, every sound somehow in the same soft register: the whisper of handsaws from the other teams, the swish of each falling branch. A little bonfire pops occasionally as the few bits of waste hazel hiss and steam.

Our task today was, until just a few decades ago, one of Britain's core industries. We're coppicing the hazel, cutting it down to use whatever products we can glean from its wood before allowing it to grow back to provide another crop. George Peterken sums it up in definitive terms: 'Most ancient woods have been managed as coppice for most of the last one thousand years.'[1] (Lots of them were coppiced for thousands of years before that, too, but the wholesale conversion of woods to a formal regime of coppicing began around the time of the Norman Conquest.) This industrial but sustainable system had the unintentional side-effect of replicating the more natural conditions of the earlier British landscape, allowing woodland species to thrive in the varied conditions in which they evolved. Here at Spring Park, the coppice regime has been restored to bring more wildlife back to the wood, as well as to renew the cultural connection that would have previously seen the place alive with human activity.

Our predecessors coppiced trees to provide innumerable essential products, from firewood and charcoal to fencing, furniture and 'treen', the collective name for the small everyday items that could be whittled or turned from wood, like plates, bowls, boxes and cutlery. Here, and at other revived coppicewoods across Britain, modern woodlanders are rediscovering these ancient crafts as they bring their woods back to life.

There are ten volunteers out with me today, a couple working alone and the rest in pairs. We're cutting a rough square of woodland about fifty metres along each side, home to a few dozen hazel bushes. The worksite is bounded by paths along the top and down one edge, while its other limits are marked with scraps of yellow rope tied loosely around thin trunks and branches. Hazel is an understorey species, growing beneath the canopy of bigger trees like oak and beech. It's naturally multi-stemmed, so even where it grows without human influence it might have a dozen or more trunks growing from the same stump, ranging in size from pencil-thin whips to gnarly chunks the width of my thigh. Coppiced hazel, like this, was usually harvested every six to ten years, with the bushes no more than five or six metres tall and the trunks at their most useful and easy to work. Hazel doesn't get too much taller even if left to grow; the trunks and branches just get fatter and the crown ever wider and more knotted.

There's a whole language of coppicing that would have been familiar even to people of my grandparents' generation who didn't work the woods, such was its importance in providing for everyone's needs. When brought into the coppice cycle, the bushes – or more accurately their stumps, which persist despite us cutting the rest – become 'stools', while each stem or trunk becomes a 'pole'. The material cut from coppice is 'underwood', or more often just 'wood' – in contrast to timber, which refers to the planks and beams that come from much bigger trees grown to maturity. Our worksite is, in the south-east, a 'coupe', exposing the Norman influence on the area: elsewhere in the south you sometimes hear 'cant' or 'panel', and in the Midlands a 'fall', while in the north it could be an Anglo-Saxon 'hagg'.

Spring Park is, in fact, named for the new growth that 'springs' back from the cut stools and is itself sometimes called 'the spring'; it's a common name for old woods that quickly indicates their working history. This Spring Park represents an intact remnant of the traditional system, with hazel coppice beneath oak standards forming one of the most widespread – even archetypal – forms of woodland management in the lowlands of Britain for the last few hundred years. The preponderance of hazel and oak is a result of both species' ecological adaptation to a wide range of British conditions, but also to the increasing commercialisation of coppice from the eighteenth century with a subsequent consolidation of the products made from any one woodland. As one of the easiest species to coppice and transport, and with a wide range of ongoing uses despite the increased availability of coal, hazel was encouraged at the expense of other trees.

The hazel- and oak-dominated lower slope of Spring Park is mostly divided into similar blocks by a grid pattern of paths; there are sixteen coupes in total, with two now cut each year. The coupe due for cutting never sits next to a coupe that was harvested the previous year, so freshly cut open areas sit next to dense, shrubby hazel that's growing back, or tall areas that might be ready for cutting the next year, the whole a constantly changing patchwork of different stages of growth and of light and shade. This 'cycle' is the key to the system's ability to provide an annual crop of essential wood, as well as the reason it creates exceptional woodland habitat.

On a clear winter's day like this, with the trees bare, we can see hundreds of metres through the wood. Looking across the slope, the understorey is dominated by the faintly geometric pattern where the hazel stools interlock and overlap – each a loose, inverse triangle of stems emerging up and out from their shared stump – with a smattering of holly and hawthorn among them. Monochrome birches, the smooth stems of young ash and the occasional hoary cherry emerge from the hazel canopy, aspiring to join the framework of big oaks above. Even within the boundary of our coupe there are a few middle-aged ash emerging into view as the hazel is cut, a clump of birch down towards a ditch in the bottom corner, and one towering oak near the path junction that forms a cornerstone of the scene and was an obvious spot for us to build our temporary camp.

They're all species that thrive here naturally in the acid soils of this gravel ridge at the edge of the North Downs, the substrate of which is easily revealed by scuffing the leaf litter. The smooth, rounded pebbles could be a river-bed and it's easy to imagine them beneath the tropical waters of an estuary, where they were laid down tens of millions of years ago. A more recent tide crashes at the north edge of the wood, where big 1930s houses face the trees across a quiet road and form the limit of Bromley's suburbs.

The edge is marked by a wide strip of sweet chestnut coppice that indicates the layers of history even within the wood. When the natural species composition was tweaked in favour of hazel and oak across the rest, the original trees were cleared entirely here to make way for the chestnut. Economic changes meant that growing longer poles of chestnut for fencing and to support hop-vines on the farms of Kent made the most sense for woodland owners, a story that was repeated across the southeast. The management of sweet chestnut, mainly for fencing, is the only type of coppicing that's continued throughout the twentieth century at any sort of scale, and still forms a strong local industry. There's a different atmosphere up among the chestnut than in the hazel on the lower slope; there are fewer big oaks, for one thing, which creates an airier and more open feel when the coupes are fresh-cut, while the longer cutting cycle of fifteen or twenty years to grow bigger poles means the older coupes are taller and more dense and shady.

The slightly jarring juxtaposition of the woodland here with the big houses over the road is no coincidence, Spring Park's very existence being defined by its proximity to London. Now a relatively small wood of only 20 hectares in size, it's a fragment of what was once a landscape of woodland across the edge of this part of South London. It's startling to look at the 1871 OS map and see Spring Park stretching north and west across Addington Hills almost to Croydon: the grid pattern of coppice compartments is actually illustrated on the old map in one area now covered by houses. Woods were necessary in proportion to the size of the human populations they served, so the ever-expanding city of London required huge resources like this to heat its houses, fuel its ovens and industries, and supply its commodities. Oliver Rackham discovered that, even in the seventeenth century, the price of wood from counties near London was almost twice the national average, due to the higher demand.[2]

To look at it on a map now, Spring Park is a textbook example of an ancient wood in lowland England, a much smaller, irregular rectangle spread along a hillside between Bromley and Croydon, and its combination of hazel with oak standards – and that more recent area of overlaid sweet chestnut – also make it fairly typical. It was forced back, as we saw at Eddie's Wood, to where the soil is thinner and, in this case, harder to both plough and build houses. While the great wood of which it formed a part survived until the end of the nineteenth century because of its utility, the combination of the rapidly declining need for coppice products and the expansion of London's population saw most of it sold for housing in the 1920s by the Lennards, lords of the nearby manor. What's left still exists not because of its use as a practical resource, but as a result of the Victorian preservation movement that recognised the importance of open space to people's wellbeing. Beginning in the 1870s with Epping Forest, the City of London Corporation agreed to own and manage land for the benefit of local people and this ancient – and, it must be said, somewhat arcane – institution slowly began to acquire sites within 25 miles of its heartland in the Square Mile. The Lennards bequeathed the remnants of Spring Park to the City's care in 1926, as the Corporation already looked after a series of old commons nearby. The commercial management of the wood was abandoned – although it received a final cut during the Second World War, as we shall see – and it was left alone as a place 'for the quiet recreation and enjoyment of the public and to preserve the natural aspect', as the Corporation's Open Spaces Act puts it.

*

Spring Park is an ideal beginner's wood, and it's where I came of age as a woodlander after a couple of earlier ranger jobs that had taught me a lot – more than I can begin to recall, in fact, as I was starting from scratch – but which hadn't involved the traditional management of ancient woods. When I finally decided to take the plunge and give conservation a go, I started volunteering with the National Trust on sites across Cheshire, simply because one of the rangers, Darren Evans, had advertised for help and they were a conservation organisation I knew from seeing their distinctive metal signs in the Peak District. I'd continued to harbour a distant daydream of woodland work through my teenage years but had stayed on the academic track that seemed laid out for me and that ultimately saw me studying American history and literature at university.

I spent a year on exchange at the University of California in Berkeley, just over the bay from San Francisco, which was where those hazy woodland ideas began to solidify; there were courses that combined literature with environmentalism, taught by a couple of inspirational professors, and my housemate and friend Nels took me on revelatory trips to the redwood groves and beyond, immersing me in the reality of the landscapes and ecosystems about which I was reading. A couple of years later, I went back to California to visit Nels and he introduced me to a passionate band of activists campaigning to save the old-growth redwood forests. Their inspiration was the final push I needed. While the realisation had been slowly dawning that I wanted to be in the woods and wanted, in the vaguest and most naïve terms, to save the planet, now I was committed to doing it – I just had to figure out what it meant in Britain.

I lived with my parents, worked late-night shifts in a big bookshop and spent the rest of my time on a desperately steep learning curve with Darren. He ignored my complete lack of practical experience and invested more effort than I could ever have expected giving me a solid grounding in the basic skills I'd need if I wanted to be a ranger; we dug leg-deep holes for signposts in the woods of Alderley Edge and then packed them in so the earth was like concrete, repaired stiles on the heathland at Bosley Cloud, strung fences and laid paths.

I learned a whole new vocabulary of tools and techniques, becoming fluent in the language of monkey-strainers and shove-holers, and he gave me the wildly exciting chance to drive the racing green National Trust Land Rover and to try loading stone with the bucket of the tractor. We spent weeks, on and off, building a big new toolshed at the yard, my clumsiness with the saw, drill and hammer gradually developing into something approaching competence. Recognising my matching ecological ignorance, Darren challenged me to identify the trees we passed, and the landscape around me seemed to emerge into focus from an amorphous green mass. My shoulders grew wider along with my perspective; I'd never been stronger and healthier, and I felt like my nascent understanding of the British countryside was something obvious and essential I'd been missing my entire life.

After a few months I was fortunate to get a place on the Trust's apprenticeship scheme for rangers, and spent time at an agricultural college while continuing to work alongside Darren and the others in Cheshire. With the solid foundation of the training scheme I moved into my first real job at Dunham Massey, just up the road in Altrincham. At this eighteenth-century deer park and historic estate, I consolidated my practical skills and began to fall in love with Britain's oldest trees: Dunham is home to a major collection of aging oak, beech and sweet chestnut and is recognised for the invertebrate fauna associated with them. The flipside to these wonderfully decrepit trees, especially at a busy park like Dunham, is that they require particular care to ensure people's safety beneath them. I also began my journey into arboriculture – the care of individual trees – as I learned to inspect them closely for signs of potential failure, as well as hauling myself shakily into their canopies to carry out pruning work.

Despite having come to love Dunham and the team there, a whole range of factors – not least the fact that lots of my friends in Manchester were moving to the capital for the next steps in their careers – then combined that spurred me to move on and I managed to get a ranger role with the City of London Corporation, helping to look after their handful of sites across the south of the city. The bunker-like modern workshop at Spring Park became my base and the wood provided my first real introduction to the tradition of coppicing.

While I – along with most of the volunteers coppicing at Spring Park today – cut my hazel stools with a pruning saw, Rob, who cuts a bit of coppice elsewhere for cash, cuts with a billhook. The billhook is not dissimilar to a machete, a flat, curved blade about the length of my forearm with a hooked point like the bill of a cartoon toucan: this is called the bill or beak. Designed to protect the main blade from touching the earth and becoming dulled, the bill also serves as extra leverage when splitting wood and can be used to hook bramble and brash away from around a stool or the base of a hedge. Lots of coppice-workers and hedge-layers are obsessed with their billhooks and there's a lively trade in antique tools that were forged from better steel and hold an edge longer than modern versions. There are also minor differences in design between regions, including double-handed and double-bladed tools, and as recently as the 1950s most workers kept different patterns for cutting, cleaving and trimming; today, the reasons for many of these variations are lost.

The fundamental design of the billhook is timeless, though, with tools that look strikingly like the one Rob's using having been unearthed from Bronze and Iron Age settlements – although the Romans are credited with popularising their use in Britain with the introduction of higher-quality iron. They were also remarkably widespread around the globe, with something similar found across every culture worldwide with a tradition of metalworking. Their ubiquity is a telling sign that their design can't be improved, but Rob still demonstrates incredible skill in cutting with his; the dense growth of poles at the base of the stool means he has to figure out just the right order of work to make space for the backswing of each short, accurate blow. Even he resorts to the pruning saw for the biggest stems – 'I can do them with a billhook but it's a waste of time' – and tells me he only really uses it to cut the young, clean growth of coppice that is 'in rotation': stools that have been cut within the last few years.

'Overstood' or neglected coppice, that hasn't been cut for many years, is too tangled and with too many fatter stems to use a billhook and even the purists tackle it with a saw. Although the coppice at Spring Park fell into neglect decades ago, the Corporation restored the regime in the early 2000s for both its cultural significance and its value for nature. I have cut plenty of neglected coppice and it is an altogether different job; it's normally easy enough to cut the chunky poles with a chainsaw, but the real time and effort is in pulling apart the big, twisted stems and their tangled crowns. We're re-cutting this coupe for the first time, eight years on from its restoration, and the poles are straighter and infinitely more manageable.

I'm too impatient to fell with a billhook – and too conscious of the rest of our busy winter work programme – but I'm envious of Rob's skill. Cutting hazel with a billhook connects our work with a thread that runs throughout human history, beginning with the makers and users of those earliest metal tools. As I look at Rob I think that, apart from his bright blue fleece and faded baseball cap, I could be looking at a Bronze Age proto-farmer or medieval woodcutter doing the same job with the same tool; and they, in turn, would recognise what we're up to and the woodland scene around us.

Although it's understood that early Neolithic people processed timber to make shelters and probably for much else, they would have done so using flint axes and wedges made from antlers to cleave bigger trees into useful sizes. Before people figured out how to make bronze, though, they learned to tend the woods and to harvest more manageable wood by coppicing, using the evolutionary ability of most native tree species to spring back to life once felled from dormant buds beneath the bark of the stump. The subsequent multi-stemmed regrowth is straight and vigorous, supercharged by the extensive root system still intact below ground: the new shoots of hazel can grow taller than me in the first summer after cutting, willow twice that. The numerous, narrow poles provided a rapid crop of useful material and were much easier to cut than mature trees, particularly considering the rudimentary tools available to the first woodlanders.

Wooden structures preserved in bogs have enabled archaeologists to date the earliest evidence of coppicing in Britain precisely. The 'Sweet Track' is a narrow walkway made from coppiced wood of a wide range of species that was dated to 3800 BCE upon its discovery beneath the Somerset Levels in 1970, making it nearly 6,000 years old. By the end of the Bronze Age, coppicing and woodland management had become sophisticated enough to enable the construction of an artificial wooden island at Flag Fen, near Peterborough, built from more than 4 million individual pieces of wood, many of them from younger trees that seem likely to have been purposely tended.[3]

Coppicing was sufficiently formalised by the Roman settlement of Britain to have a consistent name – *silva caedua* or 'cut wood' – and was regularly referred to in estate records.[4] After a likely decline during the less-well-recorded Early Middle Ages, the practice grew again in significance from around the time of the Norman Conquest and into the Late Middle Ages. By this point *most* woods in an increasingly managed landscape were under regimes of intensive, sustainable cutting and regeneration that produced almost everything people needed. Coppicewoods were routinely split into regular compartments separated by banks and tracks, and were increasingly fenced out from the wider landscape to protect the precious regrowth from livestock.

While Rob and I will tonight return to our houses of brick to cook dinner on gas hobs using metal pans and industrially fired crockery, our billhook-wielding forebears would have lived a life built almost entirely from wood, whether they were cutting five millennia ago or, in some cases, just a century. Until the Industrial Revolution the majority of houses were built from wood, particularly in the lowlands where stone was less readily available, while furniture was bodged of coppiced beech and ash, and plates, bowls, spoons and other implements – the treen – would have come from coppiced lime and maple, among other species. Carts and boats were skilfully made using a wide range of trees that best suited each component, while firewood and charcoal were integral to both domestic life and a range of industries. At points in the eighteenth and nineteenth centuries, the highest-value crop from coppiced oakwoods was bark stripped for use in the foul brew that tanned cowhides into leather.

In many woods throughout this entire period, some trees were grown to maturity among and above the coppice to provide bigger timber (as opposed to the underwood provided by coppice); Spring Park's big oaks are such 'standards'. While standard oaks are famed for providing the timber for Britain's navies and the roofs of cathedrals, at times the underwood held both greater value and more everyday relevance for ordinary people. The innate knowledge that our ancestors developed and inherited, to cut both the understorey and the standards in such a way that they produced useful things and grew back again, came, from the late Middle Ages, to be known as 'woodmanship'.

Once I've felled an entire stool into a drift of bushy stems – called, in its raw form, the 'slop' – it's time to use my billhook to 'work up' the cut poles, breaking each one down into products. It takes hard-earned intuition to heft a bushy pole in your left hand, its length running back past your right thigh, and to visualise, like a diagram of a cow with the joints of meat marked on, where to cut to make the most useful items. The easiest to identify are binders, formed only of those thinnest rods I cut from the stool first. As long as they're straight and longer than I am tall, we'll weave them in at the top of freshly laid hedges to create the ornate finish to the 'Midlands style' of hedging; all they need is the top and any rogue twigs slicing off. (Hedge-laying is another ancient craft, where old hedges that are becoming gappy at the bottom are 'laid' to restore an impenetrable barrier for livestock. Each stem is partially severed so it stays alive as it's bent over, these 'pleachers' thickening out to further increase its effectiveness. Many professional hedge-layers today use the first part of the winter to coppice hazel for the materials they need; others buy them from full-time coppice-workers, for whom hedge-layers can form an important market.)

Next come stakes, which will be driven through the pleachers of a laid hedge at a rate of four to a metre, so we need a lot. They have to be straight, thick enough to knock in with a mell, the hedge-layer's broad-headed hammer, and about a metre and half in length; I use the base of my ribs as a rough gauge of where to cut when I stand a straight length on the ground. Thinner but lengthier stems will go to vegetable gardeners as beanpoles, while any flat, pleasingly symmetrical sprays of twigs work for training peas. As the job goes on, my analysis of each bit of slop gets quicker and less conscious, and I trim side branches and sever beanpoles to length with chopping blows of the billhook, working past my leg on the far side of each stem. You can tell someone's experience by the size of the pegs they leave after cutting the branches: Rob's cuts always run flush with the stem. Mine, I think it's fair to say, are variable. A sharp billhook is a work of art in itself and will slice effortlessly through an inch-thick stem if you catch it along the grain, revealing a stretched oval of shining white wood against the speckled bark.

The rest of us periodically call across to Rob as we process our wood, checking in on the suitability of a piece for the use we have in mind as we hold it up for inspection: 'Stake, Rob?'

'Nope, too crooked – firewood!'

We make our own neat piles of different products before consolidating them each time we stop for a break; our firewood pile is always the tallest, made up of the bigger stuff that's too fat, short or bent to be a stake or beanpole. Historically, I'm sure that the proportion of each coppice-crop that went for firewood must have been much smaller, given the infinitely wider range of uses our ancestors had for wood products and the extraordinary expertise of the woodlanders to grow and process wood without waste.

It remains striking, though, to consider just how much firewood Britain needed until coal mining developed to industrial levels. The previously perpetual demand for woodfuel finally crashed between 1840 and 1870 with the construction of the railways and the transportation of affordable coal throughout Britain. Prior to that, every hearth required a sustainable supply of seasoned wood to heat the house and for cooking, while every baker and forge would have needed even more. It's sobering to think what a constant preoccupation the harvest, drying, cutting and use of firewood must have been before coal, gas and electrification, a daily exercise in planning months and years into the future.

In *The Wood Age*, Roland Ennos quotes the work of Paul Warde of the University of East Anglia, who estimated that in 1650 the British and Welsh population burned around 1.2 million tons of firewood per year; as Ennos points out, however, while this sounds like a lot – it is a lot – each hectare of coppice can sustainably produce at least five or six tons of wood each year, so it would have required only 240,000 hectares of woodland in good management, 1.6 per cent of the land area. British woodland cover has never fallen below 5 per cent (and this after coal came to dominate), so there was plenty of wood available each year for fuel, and for everything else people needed, without degrading the area of woodland.[5] Oliver Rackham, impatient with historians who wrongly blame the growth of wood-burning industries for the destruction of woodland, wrote that 'a wood need no more be destroyed by felling than a meadow is destroyed by cutting a crop of hay'.[6]

Some of the volunteers take home small loads of the firewood produced at Spring Park as part-payment for their contribution to what is, essentially, a conservation project. When I lived in a tied house with a wood stove, I'd take a trailer-full home once or twice a year, product of all our winter work across the commons, but I liked the wood from our coppicing projects best. There is a well-known Scandinavian saying that firewood heats you three times: on felling, splitting and then burning. I came to find that each stage of the process also bound me closer to the woodland itself, and to the physical acts of caring for it. On cleaving each chunk with my axe, I recalled the coupe, and sometimes the tree, that it came from, and I recognised particular lumps with something approaching nostalgic affection as I came across familiar seams in the woodpile and placed them carefully in the stove, often sitting by it with the door open to watch their peeling bark catch a flicker of flame. Hazel became my favourite firewood for its connection back to particular coppicing jobs as well as for its ease of use: it burns well, with a sweet, appetising smell, but often only needs splitting once, if at all, and flies apart readily beneath the axe.

Even the hazel twigs would have traditionally been gathered and bound into faggots, tight bundles that burned particularly hot and often fuelled bakers' ovens, or could be used as 'roading' to fill potholes in cart tracks. Some modern coppice-workers still find a market for faggots, which can be used as fascines to revet the eroding banks of rivers, or cut into short sections as 'pimps', natural firelighters. We throw a few of our tops on the bonfire, which is more for atmosphere than waste disposal, and stack the rest into 'habitat piles'; they'll make a ready shelter for small mammals straight away, then be colonised by spiders and other invertebrates, before the dead wood is consumed by fungi into the woodland soil over the course of a few years.

*

There's always a surplus of firewood from Spring Park, so between Rob and the rangers we use this to create another fundamental coppice product: charcoal. It was historically produced by 'wood colliers' beneath mounds of earthen sods, but we use a kiln, a ring of rusted steel a couple of metres in diameter and half as tall that's rolled laboriously into the previous winter's coupe in early summer, then dropped flat. Essential supplies are ferried in too: a big old canvas tent, camping chairs, and sausages and potatoes for cooking in the embers of the campfire.

Rob's in charge and supervises the loading, neatly stacking chunks of firewood in concentric rings around a 'hovel' of kindling and waste from a previous burn. Once the kiln is tightly full, we manoeuvre the heavy steel lid on top, and Rob stuffs a burning rag through one of the vents to light the hovel. There are four vents around the base, each with an opening at the bottom and a six-foot chimney, and when the wood has been roaring for around half an hour – the 'freeburn' – the vents are filled with earth to starve the flames of oxygen. This reduces the fire to a smoulder that's hot enough to evaporate the moisture from the wood and burn off volatile compounds, but which leaves the solid carbon at its core intact.

The muted burn ticks along for hours, typically overnight. The volunteers drink beer and marinate in the smoke; some bed down in the tent, while others stumble home. Rob stays mostly awake, checking the kiln for hotspots and the chimneys for blue smoke that would indicate too vigorous a blaze; white smoke – mostly steam – is a good sign, and with luck this will fade to almost nothing except the hot distortion of the air. It feels elemental to wake to the sound of a tawny owl and the smell of woodsmoke, engage in a few words with whoever else is up and watch Rob busy himself by the light of his headtorch before settling back down. In the dawn light, what passes for smoke is monitored constantly until Rob makes his judgement that the kiln is ready to be closed down, the chimneys capped and every opening smothered even more tightly to end the burn.

The following day everyone reconvenes and the cooled kiln is opened with trepidation: it's only now that we'll know for sure whether we've been successful in producing charcoal, or whether too much of the carbon burned away. A couple of the hardier helpers don masks and overalls and climb with relish into the kiln, shovelling out the blackened lumps in clouds of dust and chucking out bigger nuggets by hand. Knowing how strenuously we loaded the kiln, it's surreal to see how light the charcoal is and how quickly the kiln empties; it can be as little as a fifth of the weight of wood. Everyone else sifts it through big sieves of wide mesh, pouring the chunks into thick paper bags and letting the dust and 'fines' fall away.

There's a characteristic aesthetic to the end of a burn, when we're packing the tent back into the truck with the bags of coal. Everyone has a clean circle around their mouth and nose from their dust-mask, with the rest of their face, hair and clothes blackened except where runnels of sweat draw lines down their foreheads and cheeks. Rob sells the bags locally for barbecuing, having paid the Corporation for the firewood, but we recognise that this occasional enterprise relies on goodwill and wouldn't turn a profit if we really had to account for everyone's time. Some woodlanders do make a useful sideline in charcoal if they can regularly get a number of kilns on the go and use the down time during burns to make other products, and the atmospheric operation itself remains a good way to engage the wider public with the tradition and benefits of good woodland management.

Throughout history, though, making charcoal was far from a niche pursuit and instead a foundation of technological progress, and British woods were devoted to its production at an industrial scale; in many ancient woods it's possible to find platforms levelled out of the hillside where burns took place. Known as charcoal hearths or pitsteads, they're typically noticeable only with care, but they form vital evidence of the central importance of woodland to human culture. Stripped of moisture, the residual carbon burns hotter and with more consistency than firewood, so charcoal directly enabled the smelting of metals, beginning with copper 7,000 years ago, its melding with tin to form bronze, then iron and early forms of steel. To smelt each ton of iron, medieval ironworks needed at least 30 tons of firewood converted to charcoal, which drove the intensive management of surrounding woods and often ensured their survival.[7]

The development of effective metal tools formed a virtuous circle with the evolution of wood products, the manufacture of the tools relying on charcoal from sustainable woodland management but enabling in turn the more efficient and sophisticated felling and processing of trees, which then paved the way for the construction of some of the cornerstones of European and Asian civilisations: solid wooden houses, grand religious and governmental buildings, and ships that could sail between continents. Charcoal was also integral to making glass and strong, glazed pottery until the development of coke in the seventeenth century made these processes more efficient. (Coke is essentially the same idea but with coal as the raw material rather than wood.) While at Spring Park we sort the best lumps for sale to Sunday chefs, in the past every grade would have been put to good use, with the fines going for 'smiths' charcoal' for everyday use by the village blacksmith and, depending on species, the dust going off for gunpowder and explosives, fertiliser and – well into the second half of the twentieth century – the refinement of sugar and penicillin.[8] In the south-west of England a vital use of charcoal was in baking clotted cream.

It takes the volunteers and a couple of rangers about a week to cut each of the two hazel coupes that are due each winter at Spring Park. Mid-afternoon, as the wood begins to darken, we count and tie bundles of stakes, binders, beanpoles and peasticks and carry them out on our shoulders to stack at the yard; we'll bring the truck in for the firewood when the ground is firm, either with frost or the arrival of spring. The beanpoles and pea-sticks go to the local allotment society, where gardeners buy them for a couple of pounds each.

Professional coppice-workers continue to find more avenues for sale than we bother with. In the summer many of them use long poles slightly fatter than binders to weave 'hurdles', fencing panels a couple of metres long. Each panel is made up of nine or ten stakes or 'sails', held upright in a jig on the woodland floor, between which thin rods are woven, having been cleft in two with a billhook. Cleaving the long, narrow rods – here known as withies or weavers – is an extraordinary skill that is learned entirely in subtle and nameless changes of pressure in the hands; experienced hurdle-makers twist their billhook in just such a way to keep the developing crack even and stop it 'running' out to one side, sometimes quickly swapping hands or flipping the wood and tool to draw the force back in the right direction.

Brian Williamson has brought the old coppice at Westonbirt Arboretum back to a condition that wouldn't have looked out of place in the nineteenth century, and described to me the variations in bark colour that indicate to him whether a rod will split smoothly and take the sharp turn around the outer stakes of a hurdle without cracking; he prefers rods with a silvery sheen to the more brittle reddish sticks. Coppice craft is full of such details, minute and barely perceptible but with considerable significance for the ease and quality of the job, and it's intriguing to think how much of this knowledge has been lost even as people like Brian develop their own deep expertise.

Hurdles today go to wealthy gardeners as decorative features, but historically they were another key product of the hazel woods: before the creation of steel stock netting they were essential on farms, where they were used to pen sheep, their portability allowing for the movement of temporary overnight folds across the fells or downs, or to tightly control grazing on the freshest, richest grass before moving the flock on. Their agricultural fate was sealed when the development of artificial fertiliser meant there was no longer a need to hold livestock temporarily in small areas to enrich the soil by manuring; this practice has returned as 'mob grazing' in regenerative farming, but the price of hurdles in comparison to electric fencing prevents their reintroduction.

The flexible withies of hazel – along with other species – also formed the 'wattle' in wattle and daub walls, woven between the timber frames before being daubed with a weatherproof plaster of clay, manure, horsehair, animal blood and straw. Like many of the key coppice-crafts, the use of wattle and daub can be traced back to the Neolithic period, and it persisted in Britain until the eighteenth century and the increasing availability of stone and brick. Withies were also used to weave many other things throughout the same long period, from baskets to fish traps that spanned rivers.

I remember thinking when I first cut coppice that a finished coupe looked bare and austere, and we know from the grumpy feedback of occasional visitors that it can appear to the uninitiated like brutal clearance, vandalism of a seemingly wild wood. But to me, now, there is real beauty in the clusters of stubs with their pale cuts and in the intricate, sculptural character of the old stools themselves, dotted across the coupe and brought to the centre of our attention without their crowns – as well as in knowing that we are replicating a scene that was found every year, in every wood, from the Bronze Age until after my grandparents were born. The stools themselves are living archaeology, being potentially the oldest things in the wood as they spread and regenerate with each cut. A coppiced lime-stool in ancient woodland at Westonbirt could be one of the oldest trees in Britain; still cut on a regular cycle, the connected stumps of this single organism form a clump more than twenty metres in diameter.

Throughout much of history, the freshly cut coupe would have been fenced – perhaps with hurdles produced on site – to stop the livestock that roamed the woods eating the succulent new shoots (a transgression that could be punished by a fine in the medieval period as 'waste'). The animals were allowed back in once the regrowth was well established to browse on side branches and ground vegetation, forcing the coppice-growth even straighter. In many places today, with woods and fields permanently separated, and sheep and cows kept to their pasture, wild deer are a bigger problem and many woodlanders string black plastic netting around their coupes, using stakes cut on site as posts; it's a measure of the artificially low cost of plastic in comparison to the expense of hurdle-making that this is an unavoidable choice. The regrowth at Spring Park doesn't seem to suffer from the attentions of deer, despite us occasionally spotting roe and fallow in nearby fields, and we assume the hundreds of dogs walked here every day keep them away.

The work of a coupe isn't quite finished with the hazel cut: fellow ranger Baz and I will come back with our chainsaws to drop a few of the maturing trees of ash and birch that have grown above the coppice. The volunteers move on to laying a section of old hedge at another of the Corporation's sites up the road, entwining themselves further with the history of these places as they knock in the stakes we've cut here and weave binders along their tops to finish the job. We'll all return in spring to see the coupe begin to recover, finding reasons to pass by and observe that the wood anemones and celandines are denser than usual, the bluebells that follow thicker and glimmering more brightly; in both cases, the evidence from research elsewhere tells us it's probably true, but I don't yet quite trust my ability to observe the nuance. And then, less showy than the flowers but more exciting for it, we'll wade through the fresh bramble to examine the stools, looking for the swellings and bumps that mark dormant buds beneath the bark bursting into life: the 'spring' that will, impossibly, become whips as tall as we are by the end of the summer.

The air will be full of insects, from the easily dismissible clouds of gnats to big, gliding butterflies, while blackcaps, willow warblers and chiffchaffs will arrive to feed in our warm new clearing and to nest in the adjacent thickets of middle-aged hazel. But as well as bringing wildlife back to Spring Park, the restoration of the coppice regime has restored its human life. There is a sense of dynamism and energy from a small community of people engaged with an act of creation even as they cut the hazel down: forging the conditions for woodland wildlife to thrive, making products that transport the life of the wood beyond its boundaries and even into our homes, and sustaining – and being sustained by – a timeless tradition with meaning ingrained in the human story.

CHAPTER THREE

Being Cruel to Be Kind

How felling trees helps wildlife

THE TREE DROPS with a whump and is rustling into stillness on the woodland floor when I become aware of an irritated voice despite my ear defenders. I look up from the birch-stump already oozing thin sap, its outline sinuous around buttresses and deep fluting.

'I thought you were meant to be saving trees, not chopping them down!'

I flick off my idling saw and walk over to the plastic tape blocking the path along the edge of the coupe. It's one of the regular dog-walkers, which is always more confusing: you'd think they'd be used to it by now.

'I know it seems weird, but in a wood like this it's better for nature if there are some bits that are warmer and lighter—' I begin my regular spiel, but I can see she's already impatient to go.

'Yes, I read the signs,' she cuts in, on the move after her

Labrador, 'but I just don't see how cutting a lovely wood down helps *anything*.'

With that, she's off. I've learned by now when to persevere with a conversation like this and when to give up; I give Baz, paused in his work, a shrug and head back in to clear up the birch. I sympathise with the dog-walker. The woods that most of us have grown up knowing seem ageless and permanent: they're dense and shady places with a sense of the wild about them, as if untouched by humans. Then people like me come along with roaring saws to fell trees – when we're always told we need more of them – and create brutal clearings in the midst of this constant, peaceful refuge.

When I started out as a ranger I too had loosely understood that conservation meant saving wild places, and I didn't realise then that cutting trees down in Britain's ancient woodland – the right trees, that is – often massively improves its value as habitat, allowing nature to thrive. It was a long journey from my experience in the redwood forests of America's west coast to beginning to grasp the importance of woodmanship for British wildlife. In California, I'd finally been inspired to make the leap into conservation after meeting a group of daring activists who'd dedicated their lives to saving a very different kind of ancient woodland from unsustainable felling and use.

I awoke on a near-empty Greyhound bus, my forehead sore against the window. It had been a sleepless journey through the previous night and day, crossing western Canada and the American border on a coach packed with what I had come to learn were typically garrulous characters. I'd changed buses in Portland at midnight and drifted blissfully off across two seats as we headed south in the darkness. Opening my eyes, I saw first the Pacific filling the windows on my right, dull emerald in the dawn. In the other direction the rising sun dazzled me as it streamed through dream-like groves of towering redwoods, making me catch my breath with the startling beauty I had almost magically arrived into, as well as the excitement of being back in California. As the bus wound under the trees and past log cabins selling chainsaw-sculpted grizzly bears, it felt, in a funny way, like coming home.

The year I'd spent on academic exchange in Berkeley a couple of years previously had started me thinking that some sort of environmental work could be a reality after all my daydreaming. I studied inspirational, multidisciplinary courses about the water crisis in the American west and the idea of the wild in literature that captured my imagination and galvanised my desire to act, as well as helping me to realise that environmentalism is as much about culture and philosophy as the science subjects to which it had been restricted at school – and which had never come easily to me. I'd also been introduced to writers of ecology and landscape – and of outdoor work – like Gary Snyder, who spoke to me like nothing before, and met friends who helped me rediscover my love for the woods as well as demonstrating an attitude to life that I'd rarely come across at home.

Chief among them was Nels, who moved into my big houseshare midway through my stay. He was something of a Renaissance man, having graduated in both computer science and English, and was wild-haired and outdoorsy as well as insatiably curious. He seemed to embody a western spirit that was perhaps the aspect of my time in California that had the biggest impact on me: I met lots of people who combined deep seriousness about their work or studies with openness, informality and a laid-back, unashamed enthusiasm that contrasted sharply with my experience of university – even life – in England. As well as joining me on trips to the absurdly picturesque skatepark on Alameda Island, with its view across to San Francisco via the Bay Bridge, Nels took me hiking in the redwood groves near Santa Cruz and in Marin County, and led a carload of us on an eventful and mind-expanding trip to Zion Canyon in Utah. Those trips further reminded me how good it felt to be out in the wild – as I thought of it then – and increased my determination to do more of it.

After a final, slightly dispiriting year of study back in England, I went to Canada for a few months to work on a ski resort. Trudging through tedious jobs in hostels and hotels for the chance to snowboard a couple of times a week clarified my existing sense that I really needed to find work that I enjoyed and that contributed something meaningful to the things I cared about. To my surprise, I also realised that I enjoyed the bits of physical work I did, shovelling snow or helping to fix doors and furniture. And the pay-off was that I had some of the most exhilarating days of my life, hiking Rocky Mountain ridges to drop down untracked gullies of powdery snow before swooping and floating through the trees below, experiences that only reinforced my desire to find work that got me outdoors and into the hills and woods.

Nels had been to film school and was embarking on a career as a documentary maker when I travelled from the Rockies to visit him before flying home. For his first film, he planned to explore both sides of the bitter debate around logging the west coast's old-growth redwoods, and I stepped off the bus to meet him and his brother Jack – always called Toots – on a deserted road in Arcata, a small town a couple of hundred miles north of San Francisco that lies at the heart of redwood country. They were on their second stint living in a caravan and following two characters whose contrasting perspectives would illustrate the story: a gruff but charismatic old logger, Bill, and a radical activist who called himself Lodgepole, after the species of pine – LP to his friends.

Arcata was almost tailor-made for my emerging interests. It's home to Humboldt State University, which, I would later discover, hosts many of the scientists who used pioneering climbing techniques to access the crowns of the redwoods and explore the unique ecosystems found there, further emphasising the ecological significance of the forests. The town was, if anything, even more alternative than Berkeley, while the university ran ground-breaking courses on sustainability and environment.

It attracted progressive students and activists as well as a wider community of people with a related spectrum of interests, from wholesome experiments in communal eco-living through to dropping out and smoking weed all day. Surrounding the town, however, were communities built on logging, which I was also fascinated by, having been captivated by Gary Snyder's Zen-inspired poems of forestry work in his collection *Rip-rap*. I was sternly warned that the loggers hated the 'hippies' and that, now associated with the town's activists, I should never try to hitchhike.

Lodgepole was involved with most of the interesting things going on in Arcata, always busy and often difficult to get hold of, to Nels' continual frustration. After one fruitless series of phone and house calls over a couple of days, we found him smoking with the dozens of perpetual stoners who inhabited the town square.

'Why didn't you just come downtown?' he asked an exasperated Nels in a caricature *Bill and Ted* California drawl: 'You'll find me.'

Lodgepole's faith in the universe proved well-founded, and thereafter if we wanted to catch up with him we'd drive into town and mooch around until either we stumbled into him or someone mentioned his whereabouts. Lodgepole had agreed to feature in the film but was still cagey, unsure of Nels' motives and how he'd be represented; his reserve extended to me as Nels' unexplained companion, and, never cool, I felt particularly square here with my English accent and perhaps the shortest haircut in town. It turned out, though, that Lodgepole was into skateboarding, and upon discovering that I was too his exuberance overcame his caution, and he took me immediately to the flowing, concrete skatepark on the edge of town. Sat on its wall a couple of hours later, sweaty and stoked, he told me with an air of generosity, 'You can call me LP.'

LP filled me in on the situation 'behind the redwood curtain', as he called it, in the rural communities of northern California.

As I had already begun to grasp, there are two species of redwood in the American west, the giant redwood and the coast redwood. The giant redwood is native to the Sierra Nevada, the mountain range inland that's home to Yosemite and Kings Canyon National Parks, and is the world's biggest tree by sheer bulk. The coast redwood is the world's tallest tree, and its range is limited to the coast of California and Oregon by its partial reliance on sea-fog to provide moisture to the tops of the trees, where the organic hydraulic systems that other species rely on to move water from the soil begin to fail. The biggest specimens of both species can live 2,000 years, making them some of the oldest organisms on earth as well as the biggest.

Unfortunately, one of the characteristics that makes the coast redwood so long-lived also made its timber irresistibly attractive to European settlers: it's rich in tannins, which, in ecological terms, prevent fungi colonising and decaying the wood. The tannins provide the same benefit once harvested, so it makes strong, long-lasting timber. With each giant tree yielding hundreds of tons of high-quality building material, the redwoods fast became a valuable resource; much of San Francisco was rebuilt with redwood timber following the 1906 earthquake and fire that destroyed the growing city. The rate of felling increased with the development of chainsaws and bulldozers to drag sections of trunk off the hills, and, by the time LP and I sat looking out on the wooded hills from the skatepark, only 5 per cent of the old-growth forest remained.

I know now that indigenous American people influenced the forests in lots of ways – by burning areas to create clearings of lush growth that attracted game, for instance. But with their lower population density and very different histories of metallurgy and building, they didn't harvest timber in anything like the same way as in the coppicewoods of pre-industrial Britain. While they haven't been truly 'wild' for millennia, the structure and dynamics of the old-growth redwood groves are much more natural than the woods of Britain and Europe; and with their old trees essentially irreplaceable, they're much better for their abundant wildlife when left alone.

As early as the beginning of the twentieth century, some areas of old-growth redwood were incorporated into state and national parks, but some remained in the hands of private landowners and forestry firms that continued to log them. Their ecological importance was only just becoming understood, even then: while second-growth redwood forests – which develop naturally where the old-growth is felled – can still be rich in wildlife, it was only with the work of Humboldt State's canopy-climbing scientists in the 1990s that the staggering ecology of the old trees was fully revealed, with aerial ecosystems of ferns and other plants growing in the crowns, a species of salamander that never descends the vertiginous trunks, and the Humboldt marten, a relation of the British pine marten that was rediscovered, having been thought extinct.

Lodgepole was involved with a loose collective of protesters who trespassed into the old-growth forests marked for felling and climbed the biggest redwoods they could find to 'tree-sit', building camps high in the crown to impede the foresters, sometimes for months or years. LP was, it transpired, a trained arborist – a tree surgeon – who used the rope skills he'd learned pruning trees to access the giants. As well as physically preventing the felling of individual specimens, the tree-sitters' camera-friendly treehouses and confrontations with loggers helped to raise the profile of more mainstream political campaigns for better protection of the forests.

I was aghast that these monumental trees were still threatened, having assumed they were all protected – especially as some of the concentrations of redwoods are iconic tourist destinations, like Muir Woods near San Francisco. At the same time, I was completely enthralled by LP's tales and wildly impressed by these passionate activists climbing the giant trees to build Ewok-style camps in their crowns. I asked LP to take me to see some old-growth forest, keen to understand more and to unpick the difference in the redwoods I'd come to love between second-growth forests and the really special trees. The following day the four of us drove a short way into the local state park, Nels and Toots leaving their camera behind at LP's request: 'I can't take you to a sit, they're secret and we might get the shit kicked out of us by the loggers – but this place is still sacred.' He led us from the road down a hairpin path into a tight valley. It was shady and dark, and looked to me like beautiful natural woodland, the trees bigger than most in the woods I knew at home and the understorey lush with ferns and shrubs.

'Second-growth,' LP asserted brusquely, striding on purposefully before pausing. 'Look.' He pointed uphill into the trees, seeming to soften, then led us off the path. We scrambled to a looming stump wider than I am tall and twice as high, like a relic from a lost civilisation that had been consumed by the surrounding forest. We picked our way up the notches struck into it with an axe where, LP told us, planks would have been wedged as working platforms for the loggers. We perched in a row on the rim of solid-ish wood around the decaying centre of the stump.

'From the age of the other trees and the height of this thing, I'd guess it must have been felled a hundred years ago or more,' LP surmised; 'They had to get up past the buttresses with the old saws. I wish they hadn't done it but you gotta give it to 'em – imagine perching on a plank up here pulling on a crosscut saw all day, then leaping away when it started to fall!'

He led us on along the main path before suddenly turning off on a barely perceptible trail between a couple of car-sized boulders, where I quickly lost my bearings as we scurried to keep up. Before long, though, we emerged into the valley floor, where LP's description of the forest as sacred was made suddenly and unarguably real. Trunks the width of the stump we'd climbed surrounded us, but instead of stopping above our heads they soared skywards, hiding one another and confusing our sense of perspective as we craned to peer vertically; we couldn't make out their tops but lost them to their shared canopy a disconcertingly long way up. They were spaced more widely than the second-growth trees, creating a soft clarity in the cool green light around us.

The ferns were thicker and more luxuriant, and LP's commentary waned as we picked our way through the denser maze of understorey shrubs. He looked back over his shoulder: 'Check this out.' The sun hit us directly for the first time as we entered a clearing, specks of dust suspended in the space above. The rootplate of a fallen redwood rose perpendicular to the ground, the time since its collapse marked by the hanging garden of ferns and creepers that dripped from its elevated roots. At its centre, gaping like the mouth of some giant but strangely benign monster, the centre of the trunk had decayed to form a cave into which we could walk for three or four metres, cool and smelling pleasantly earthy.

'There's cavities like this in the tops of some of them,' LP told us, hushed, 'they got owls living in 'em and all sorts. You wanna see how tall these trees really are?' He led us around the exposed roots and we climbed on to the soft bark of the prostrate trunk; it was laid along the earth with such solidity and with the needle-litter and understorey having consumed it to a point of such seeming permanence that it was barely credible that it had once been upright. The stem stretched off like the hull of a hairy submarine, its red bark fibrous and ridged but otherwise featureless. We followed LP in single file along it, our feet at the height of the surrounding scrub, until after thirty or forty metres we reached the first skeletal side branches. We picked our way through them until they became too dense to navigate, then dropped down to trace the last few metres on the soft duff of the forest floor. The tree seemed to stretch back for a mile, and LP smiled at our awe.

'She's only a baby – I paced it out at about eighty yards. The biggest ones we know about are more than a hundred. I've climbed a couple, I reckon.' He seemed to know his work with us was done – as if we'd needed convincing.

'This bit's safe, it's in the state park. But there's plenty of places like this they want to log. These trees been here thousands of years and those bastards fell 'em in a day! What gives them the right?' He paused, then told us with an air of nonchalance, 'This is what we're fighting for.'

Inspired by Lodgepole and his crew to finally act, on my return to England I started volunteering with the National Trust and found my way into those first couple of ranger roles, in the Cheshire countryside and at Dunham Massey. Going into it, I see now, I didn't really have a clue what conservation in Britain was all about and perhaps assumed that, like LP or similar activists in the rainforests of Brazil or Borneo, here too we'd need to save ecosystems from exploitative human attention. All I had was a strong, albeit hazy, sense of wanting to help and, from my very limited experience, I liked the idea of what a ranger did – and I knew I wanted to be among the trees.

My first jobs provided a slow start in terms of woodland management. Our work in Cheshire, while giving me a strong foundation in the practical skills I'd need, was mainly focussed on improving the network of paths and stiles as well as restoring lost heathland, while at Dunham – despite the park's status as an important historic wood pasture – our focus was on the care of individual trees. At both places, most of the felling I'd done had been to remove trees that could pose a hazard to people. Our lack of involvement with the woods reinforced my impression that they were better off left alone and would look after themselves, like the redwood groves, and that our job was about managing the interactions of visitors with the special places in our care. It was only when I came to Spring Park that the true richness of Britain's ancient woodland and the role of people in its ecological story was slowly revealed to me.

Baz Gutteridge, my boss at Spring Park, is a quiet man with a cautious approach to things, but his reserve belies a steely determination to improve the sites he looks after and he has spent twenty-five years steadily nursing the woodland here back to health. When he first started, he explained, the only work that had taken place – probably since the Second World War – had been to improve the wood's paths to ensure that local people could enjoy it, the core purpose of the City of London's ownership. Baz had successfully reintroduced coppicing to much of the wood, dividing the hazel and oak of the lower slope into a series of coupes as well as the sweet chestnut along the top. It had, though, been 'quite painful', as he put it.

Keen to avoid alarm caused by a misunderstanding of tree-felling, Baz had publicised the plans in advance, explaining the benefits of coppicing through letters to local residents. Things didn't go to plan. While lots of locals were prepared to listen and to trust the rangers, a couple of loud voices ran a campaign that painted the proposal as a cynical money-making scheme by the Corporation. He showed me a leaflet distributed by the campaigners that claimed that the supposedly clear-felled 'timber' would be shipped to Portugal at a huge profit, a bizarre accusation founded in nothing resembling fact.

Eventually, however, after several bruising public meetings and a series of articles in the local paper, the rangers had been able to slowly start cutting coupes, albeit with more scrutiny and worry than they'd have liked. By the time I arrived, everywhere in the wood that was marked for restoration had been cut once. It was exciting to help with the second cut of my first hazel coupe with the regular group of volunteers, felling the stools and – very slowly – working up the slop, relying on Rob's input to help shape almost every stake and beanpole. I had, by this point, also begun to read Oliver Rackham and was beginning to grasp the historical significance of coppicing in ancient woodland. If anything, I was more conscious of the restoration of the human tradition than its benefit for wildlife, even as Baz explained the increases we should see in butterflies and wildflowers.

Come spring, the wood anemones and lesser celandines formed a spectacle in our fresh-cut coupe, followed by the bluebells, but any difference from the rest of the wood was minor. These three flowers have evolved to use the spring sunlight that reaches the woodland floor before the trees are in leaf, allowing them to grow beneath a dense canopy. They can, therefore, persist in mature woodland, including under neglected coppice and older coupes. There are other flowers that should, in theory, materialise later in the season in fresh-cut coupes; around two-thirds of woodland flowers, in fact, respond positively to coppicing, including some whose seeds can persist in the soil for decades. But, as the months rolled on, the changes to that first coupe weren't as dramatic as I'd envisaged. The dormant hazel buds did emerge into a couple of dozen whippy new twigs on each stool, but they weren't exactly racing away; they petered out at about a metre tall by the autumn. Surrounding them, a few scrappy ropes of bramble looped their way across the leaf litter amidst solitary fronds of bracken. I didn't think too much of it; we had, after all, continued the tradition of coppicing by cutting the hazel, and perhaps all sorts of magic was happening for wildlife that I simply didn't understand. Soon, however, I would witness the true, unmistakeable explosion of life that coppice management can ignite.

As well as working at Spring Park, I help out at the City's other sites across South London. As one of the few rangers trained to climb trees, I spend a few weeks each winter at Ashtead Common, carrying out careful pruning work to prolong the life of a rare population of ancient oaks – which we'll learn more about in the next chapter. There I met Jamie Simpson, an expert contractor who is both deeply knowledgeable and incredibly passionate about helping the wildlife of the trees and woods. Jamie quickly became a mate, so when he invited me to help out at a special wood he'd mentioned it felt like a privilege despite being a weekend's unpaid work.

He picked me up early one summer morning, our saws, fuel, camping gear and other kit packed neatly in the back of his pickup, and we drove down to Dallington Forest in the High Weald of East Sussex, one of the most extensive wooded landscapes in the south-east of England. We turned onto a rough track and bumped along through miles of woodland and endless conifer plantations, past occasional gated mansions as well as quaint old cottages, gradually dropping down the side of a valley until we parked in the junction of two holloways with overstood chestnut coppice rising high above us.

Jamie had been asked by friends in London to look after Forge Wood, a small but varied ancient woodland at the heart of the forest. There are giant beech pollards surrounded by pines on lost wood pasture, as well as swathes of neglected coppice of a whole range of species, from chestnut and hazel to hornbeam and alder. Tight side-valleys – 'ghylls' in Sussex, as in the Lakes, for some reason – capture humid microclimates that support mosses and lichens more typically associated with the temperate rainforest of the west of Britain.

The sense of history at Forge Wood is even more palpable than at Spring Park. As well as the deep holloways through which the main stone track and a couple of other muddy paths run, there are dramatic ditches orientated downhill that all converge on the little hamlet of a few houses below. There had been a decent-sized forge there from at least the time of the Tudors, Jamie told me, when Sussex was home to a flourishing iron industry; it could even have been a bloomery for smelting raw iron from the deposits of ore in the wealden clay. The coppiced underwood, as well as charcoal made on level hearths cut into some of the slopes, was dragged by horse down the radiating channels to fuel the furnaces. The flat valley bottom, now a meadow that was playing host to a couple of fallow bucks, was once a 'hammer pond' that would have fed a waterwheel that powered big bellows as well as machinery for pounding the cooling metal into ingots or tools – or possibly, here, into artillery and ammunition, as a cast for cannonballs was found nearby. Places like this around Sussex are sometimes also known as hammer woods.

Forge Wood's varied tree species are also a mark of its history. Woods like Spring Park were 'modified' from the eighteenth century to their simpler system of hazel with oak standards, as the Enlightenment-inspired trend for 'improving' the land – followed by the onset of the Industrial Revolution and increasing market economy – led to a decline in the local use of whichever tree species any one place was naturally home to. It incentivised instead the rationalisation of woods into coppice systems that could more efficiently produce larger quantities of fewer products for transportation and sale elsewhere. Less useful or profitable tree species were killed or torn out and the desired trees 'promoted' by planting or by forcing vegetative reproduction (thin hazel poles were bent over and pegged into the earth, for instance, where they'd take root and grow as new stools). Coppicewoods that remain mixed, like Forge Wood, weren't modified so dramatically, probably because they were more remote from towns and transportation, and they consequently reflect a more natural species composition and an earlier phase of woodland history. That said, the presence of sweet chestnut – a non-native species planted widely in the south-east during the same period – and the separation of some species into distinct coupes are both testament to the importance of the woodland as a resource, and its sophisticated management.

After our brief tour with a cup of sweet tea, Jamie took me to our job for the day. Running along the top edge of the wood, beneath the neighbouring pine plantation, was a long, sunny hillside of hazel; in his first intervention after a couple of years spent clearing the woodland of invasive rhododendron, Jamie had coppiced the whole thing. The effect was extraordinary. There was a clearing the length of a football pitch and half as wide, with dense bushes of tight-grown hazel poles taller than us as we walked between them. Jamie grinned: 'Only two years since

I cut 'em!' Where the hazel was sparse, it was interspersed with bright purple stands of willowherb and foxglove; in other places there were violets limply going over amidst the sward beneath, and Jamie told me there had also been common spotted orchids earlier in the season. There were, I noticed, only half a dozen middle-aged standard trees rising above the wide area and casting their shade across the bank.

At the top, between the coupe and the plantation, there was a vague but unmistakeable woodbank, an ancient marker of the boundary, with a gnarly old hornbeam pollard perched on it along with a line of younger trees: more hornbeam and birch, and the odd oak.

'We're gonna pollard them, Luke, like the old boys used to do on the woodbanks, and it's gonna get a load *more* light down here!' I spent the day as Jamie's 'groundie' as he used his climbing gear to get up each of the trees on the bank and bring them down to a stub a couple of metres tall, resurrecting the tradition of the old woodlanders of leaving pollards to mark the boundaries of their coupes. Jamie works harder than anyone I've ever known and I had to race to keep up with the branches that rained down, snedding the big ones with my chainsaw and, under his friendly, urgent direction, dragging the brash into a long and sinuous habitat pile that resembled an Andy Goldsworthy installation: 'Might as well make something cool while we're at it,' said Jamie, although the whole project felt like a deeply creative act.

I looked along the woodbank with its new row of pollards. It was transformed from a subtle historical artefact, visible only to those in the know, into a meaningful, living entity, the low bank now clearly defined for anyone to see and drawing a philosophical as well as physical line between Jamie's wood and the plantation beyond. Time felt more alive, although I wasn't sure if we'd stepped back into an ancient scene – like something from a painting by Bruegel or Constable – or if we'd dragged an old landscape into the twenty-first century; either way, it felt good.

It was clear to me, by the end of that first day at Forge Wood, what we needed to do back at Spring Park: *get a load more light in*, as Jamie had said, by felling more of the canopy trees. If anything, the impact was exacerbated when we returned to his hazel coupe on a wander as the sun set. Looking down from the woodbank, the still air was thick with midges and gnats, some in columns dancing up and down, backlit by the setting sun, while clouds of them gathered chaotically elsewhere. The place was very literally humming with life. The odd bee and hoverfly still patrolled between the foxglove flowers and a red admiral butterfly quartered the scene in sporadic flight; not particularly unusual to see, but rare in woods, where it can't cope with the cool shade. Scratchy, bubbling birdsong rose from the thick hazel below us.

'Garden warbler,' said Jamie, 'so my mate Graham says – he does the bird survey here. They're back since we cut it. He's heard blackcaps too, and willow warblers.'

I returned to Spring Park with the zeal of a convert and scared the life out of Baz with my enthusiasm for taking a bolder approach. As I should have known, he was well aware of the need to get more light into our coppice coupes, but was playing the long game and planned to take the standard trees out progressively with each cut of the hazel, cautious of the response of the local residents after his tough experience when the restoration began. George Peterken wrote in 1981 that 'the significance of coppice management for ecologists can hardly be exaggerated',[1] and powerful evidence for the need to let more light in to the woods had been around since the 1990s, Baz told me, when experts like Rob Fuller of the British Trust for Ornithology and Martin Warren of Butterfly Conservation had demonstrated the link between the declines they'd observed through detailed surveys of woodland birds and butterflies with the increasing shadiness of neglected coppicewoods.[2]

Their research showed that you actually need to get canopy cover – literally the proportion of an area covered by the spreading crowns of trees – down to less than a fifth of a fresh-cut coupe in order to see measurable increases in the wildlife that needs warm, open areas in woodland: this can increase light levels twenty-fold when compared to a normal woodland canopy. And while their work focussed on birds and butterflies, noting particularly the decline in less common specialists like the nightingale and rare fritillary butterflies, more recent studies have also emphasised the benefits of coppicing for the sheer number of more common and less glamorous species, like the gnats and midges that I observed in the evening light at Forge Wood. In some respects this should be no surprise, as monitoring butterflies is a useful indicator of the health of other insect groups that are more difficult to identify and survey, but it does illustrate the importance of the sheer *quantity* of life throughout the food chain: 'bioabundance' as well as the better-known biodiversity.

It is, after all, the less charismatic insects that woodland birds feed on, so the profusion of gnats and midges – not to mention bees, flies, caterpillars, spiders, and everything else – is part of the reason they return to restored coppice; there's lots of food, and places for them to find it, but also dense cover in adjacent, older coupes where they can shelter and nest. And lots of those insects rely on an increase in plants that provide an equivalent abundance of their food (as well as simply on the humid combination of damp earth warmed by sunlight, in the case of the midges). This, in turn, helps us understand the true significance of getting enough light to the woodland floor to encourage the profusion of nectar-rich species like bramble, foxglove and willowherb. They're common enough plants of the wider countryside that it's easy for us to overlook as we search for the more unusual flowers that respond to coppicing, but, by dint of their sheer volume, they underpin the boom in life that extends up through the entire ecosystem.

Wildlife hasn't somehow adapted to this human regime in the last 6,000 years – a moment in evolutionary terms. Our ancestors managed coppice intensively to produce the things they needed, but in doing so they unintentionally sustained a rough proxy for some aspects of the wildwood, the landscape before radical human modification – as we'll see in Chapter Five. With its areas of mature trees, open glades and every stage of scrubby growth in between, coppiced woodland supports so much life because it replicates many aspects of the complex and vital combination of habitats in which all species evolved.

The very fact that the majority of native tree species are able to coppice – to spring back to life once cut – is in itself an ecological adaptation to the conditions of the wildwood: most trees evolved in the Tertiary Period, between 65 and 10 million years ago, when temperate woodland would have been crashed around, broken and browsed by megafauna: super-elephants, super-rhinoceroses and super-horses, as Rackham called them.[3] In response to this abuse, the trees adapted over generations to survive despite being bent and torn, developing a trait close to immortality in their ability to produce new growth from broken stems and stumps. When we see coppice stools bursting back into life today or, especially, when we see the exuberant, irrepressible regeneration of a collapsing lime in a lost corner of the woods, we're witnessing the continuing impact of those long-extinct herbivores.

That's not to say that coppiced woods form an exact replica of the wildwood. Coppicing has particular effects that mean its communities are unique, for better and for worse. There aren't the wider, grazed areas we find in wood pasture landscapes, for instance, with the species we mainly associate with open grass- or heathland; in a well-managed wood, though, these habitats can be provided by permanent glades and sunny 'rides', linear clearings along access routes. During the period of intensive historic coppice management, there probably wasn't as much 'old-growth' woodland, either, with its particular species dependent on decaying wood and shade. The removal of so much woody growth so regularly, on the other hand, reduced nutrient levels in the woodland soils, which probably further favoured the more delicate ground flora that would otherwise be swamped by coarse vegetation. Even now, we might actually find some woodland species, like bluebells and some of the other iconic spring flowers, in greater concentrations than would have occurred naturally, due to the exclusion of livestock from many ancient coppicewoods – although this is still within the context of their loss from most of the 97.5 per cent of Britain that isn't remnant ancient woodland, and where they would have once been found throughout immeasurably larger woody landscapes.

Baz, with eternal patience, indulged my enthusiasm and agreed to us felling more canopy trees as we coppice the hazel at Spring Park, although we're still working towards that 20 per cent cover in each coupe over a few cycles. Before we begin cutting, we agree which trees to remove from the middle-aged generation of birch, ash and cherry that have grown among the hazel since the war.

It's usually a quick conversation on the day but the decision will have been long in the making, considered carefully every time we walked past all summer and on every tea break when we were cutting the hazel. We can't just take one year's coupe into account; we need an understanding of the canopy that will remain across the whole wood, ensuring that there's a range of species and ages of tree, and preferably a range of ages of any one species. Forester and writer Robin Walter describes it as 'creating a living structure in four dimensions': the three physical dimensions plus time.[4] This is the art of woodmanship, although in some respects it's easier for us than for our ancestors as – here at least – we're not trying to make essential products from the trees we fell, or ensuring the continued provision of useful timber for our successors. It can still make your head spin, but by the time we arrive there's usually a plan and we get stuck into the wordless, immersive process of felling.

I'd be lying if I said that felling isn't really fun. It forces an intimate connection with a tree's entire physical reality, outside and in, and there's an undeniable rush as the tree starts to move with deep clicks, then gathers momentum and thumps to the ground with enough force, from a bigger specimen, that you feel it through your boots, leaving the bright new gap above filled with flying twigs and swirling leaves. The particular smell of the sawdust of whichever species you're cutting fills your nose and they all sound different in the way they creak and tick before they fall, and even in the way their crowns whistle through the woodland air. There's hard-won skill in reading the weight and lean of a tree and in accurately placing precise cuts with a roaring chainsaw so it falls just where you want it.

The ash I move on to after being sidetracked by the unhappy dog-walker has a heavy lean, so I'm particularly cautious; a forward lean can be even more dangerous than a tree that's sat back and needs winching or pulling over. The wood of ash has strong, flexible fibres running through it, which made it ideal for its traditional use as tool handles and, more recently, sports equipment like hockey sticks. The same fibres mean it's particularly prone to splitting up the stem before your cuts are finished, if the pull of gravity is too strong, so that a strip of trunk springs violently back towards your head, a mishap known as a 'barber's chair' that can be fatal if you get caught out. After a considered moment sizing things up, the point of commitment comes as I start to cut the 'gob', the wedge removed from the front of the tree. It's vital to get this right as it dictates the direction of the tree's fall; when I put the back cut in, I'll leave the strip at the point of the gob intact to act as a 'hinge' that the tree pivots over.

Although the saw is singing noisily in my hands and jumps as I squeeze the throttle, I touch it to the tree gently to ensure it's in the right place, and the smooth grey outer bark peels clear to reveal the thin layer of light green bast – the undersurface of the bark – before the blurred chain sinks into the wood. I spent the end of our lunch break sharpening my saw and the dust plumes out in curling flakes, the hooked teeth dragging the vicious little blades deeper with each rotation. A chain that's lost its edge produces powder and the smell of hot metal, and you have to push the saw through the wood.

The pale wedge zings out, propelled by the chain. I look up for Baz across the coupe, bent across the tree he felled last as he sneds it, trimming the side branches off the trunk. It doesn't take much for us to communicate what's going on, a nod from me and a quick thumbs-up from him: we've been working together a while and there's a natural rhythm where we alternate our fells and almost have a sixth sense for when the other person's ready to go. He's well clear but pauses for a moment to watch. With that, I'd normally push the saw into the back of the tree to let it fall. On this ash, with its alarming lean, I bore in from the side to reduce the chance of a barber's chair, then pause for a moment to take a breath – not to mention a little shuffle backwards on the one knee I have down – before revving hard to sever the last remaining peg so that it goes quickly without having a chance to split, the thin hinge of wood behind the gob all that's left to guide it. The tree drops in an instant, with a twang from the last tearing fibres and a shaking thud into the leaf litter as I take a couple of quick steps back and watch it settle.

That's how it happens. The experience of felling, though, and the skill that goes into doing it safely and efficiently, is much richer but exists in a million tiny details that I rarely consciously think about. It's in the patterns of moss across the bark as I lean close to sight my cuts, and in the very particular smell of weeks-old sweat and a paste of chain oil and sawdust in my protective chainsaw trousers. The thin blue exhaust that erupts from the front of the saw and lingers on my sweater. The blinkering effect of having my ear defenders on for hours at a time, lost in my thoughts one moment as I clear up, closely focussed on careful cuts the next. The countless instinctive, wordless decisions at the base of the tree around the shape of the trunk, the species, the particular crook of the branches in the crown and what they mean for its weighting; the slope, the height. It would take an encyclopaedia to describe what enables the quick, instinctive actions that I struggle to explain when asked but without which any account of how to fell a tree seems almost pointless.

There are things that Baz and I are able to verbalise as we eat our sandwiches, like how each species of tree responds differently to the saw. Some are reliable: oak and beech, as you might imagine, and, perhaps less obviously, birch. I always feel more confident that they won't do anything too unexpected when I cut them, and I can pull them around a bit away from the direction of their lean, trusting that the hinge, that last remaining strip of wood, will hold. The cherries I've felled always seem to have decay in their centre whatever their age, and the remaining wood often feels brittle and unreliable. Ash, like the one I just dropped, is one of the worst trees to fell, even when young and free from decay, with its brittle branches and springy propensity to split.

There's a perceptible difference too in the ease with which your saw chomps through the stems. The toughest wood I've cut is long-dead sweet chestnut; the softest, lime, which caught me out on my first time as I pushed the saw too far, too fast, expecting much more resistance. Limewood is noticeably less dense than most other British trees, both to lift when you're moving cut stems around and to look at on a stump. There are pronounced rings of wide xylem vessels, the tubular cells that run vertically up the trunk to transport water, easily visible as they honeycomb the wood.

Like a medical student dissecting a cadaver to learn to heal the living, felling trees has brought me a deeper understanding of the different species and their identities in a way I use in a practical sense every day, but it's also increased my appreciation and connection to them in less tangible ways. I love oak all the more for knowing the clarity of its rich, tea-stained annual rings and the starburst of vivid rays on a fresh stump; the contrast with its casing of bark, fibrous like a coconut shell in cross-section, and the sharp smell of its tannins. The cut wood of beech and ash is lighter in colour, and beech sometimes reveals irregular thin black lines almost like a map. Known as spalting, it's where particular fungi have just begun to colonise the wood; the dark borders mark where the territories of different individuals meet, and the patterned wood is prized for turning. Some decay fungi cause the softening wood of big beech trees to stink of expensive soft cheese, at best, or vomit at worst. Alder famously opens up to reveal livid orangey-yellow wood that fades in time to deep red.

And while I'm sure lots of people are attentive enough to notice the particular form of the branches and twigs of the different species of tree as they fork, droop, bend and taper, I think I learned their subtleties as much by feel as through visual observation, snedding and cross-cutting and dragging them endlessly with technique thoughtlessly tweaked to their differences; the best place to cut, the size of branch I can pull, how to hold a bundle for best grip. Dragging straight, light bunches of birch with their inescapable witch's-broom effect is very different to tackling awkward armfuls of oak branches in a tangle of elbows and knees.

I still find the responsibility of choosing which trees to fell sobering, even having spent years watching nature respond to the places we've worked. I used to be intimidated and not know where to begin: which are the right ones? I'm now comforted by the fact that there is no 'right' answer and that almost every woodlander would do things slightly differently, applying their own personality and philosophy to the process while always, in some way, replicating natural processes and following what the wood itself shows us. Of course, there are sometimes the *wrong* trees for felling, those that support a rare species or create a particular microhabitat for a wildflower or lichen; ecological monitoring is essential in informing our work.

Here at Spring Park, the grid pattern of rides, probably first installed during the eighteenth-century 'modification' of tree species, suggests the design of the coupes, while its amenity value to local people means that Baz is right to tread gently in felling bigger trees, respecting what its many visitors want from the wood even as we try to revitalise the rich ecosystems we all need, whether we know it or not. Jamie operates with more freedom over the older and less ordered terrain of Forge Wood, and with fewer people nearby to object to the bold decisions that come from both his energetic personality and his long experience of working alongside some of the south-east's remaining commercial coppice-workers.

Both ways are OK: every wood is unique and the story of each has been written by the varying needs and desires of generations of woodlanders, as well as by its geology, climate and other natural factors. The differences between sites are a good thing, as long as our individual approaches are ultimately focussed on restoring or improving the habitat value of our neglected woods.

The felling over each coupe at Spring Park doesn't take more than a couple of days and, when we stand back to check we're finished, the trees we've chosen to leave take on a new importance under the changed perspective, both visually – we see them properly for the first time – and in terms of the role they'll play in the future of the wood. It can be disorientating to consider these new standards: they take on a different meaning as they're revealed from the anonymous canopy, singled out to perform a special function. Our successors might fell them for timber as our predecessors did, carefully ensuring a continual crop of larger-diameter wood for their essential needs; here at Spring Park, the main thing they'll provide is more niches for wildlife – foraging ground for nuthatches and woodpeckers, grazing for insects – and simply the continuation of a definitively wooded scene that comforts and restores the people who visit.

Come summer, nature returns. We all agree that the bluebells and wood anemones form a thicker carpet, while yellow archangel, clinging on at the edge of Spring Park's paths, appears in bigger clumps. Slightly anonymous at first, like a little nettle, on closer inspection its yellow flowers form a cheering spectacle. We find St John's wort and dog violet, while loose stands of foxgloves appear around the edges of the coupe, forming the foundation for a web of life.

In the absence of the giant beasts that once crashed the trees of the wildwood around, we need to play their role in Britain's small, fragmented ancient woods, as our ancestors did – albeit unwittingly – throughout history. The crisis in nature creates a new need for woodland management, as essential as the needs of previous generations for firewood and charcoal; the long history of coppicing provides a vision for the future where a renewed tradition of working the woods restores their complex ecosystems along with their human life.

CHAPTER FOUR

Common People

Wood pasture, common land and ancient trees

I DO MY best to climb the ropes rather than the trunk of the oak, hauling my weight with my right hand while shuffling the locking knot up with my left and twining the line between my boots for a bit of extra impetus. I want to avoid scuffing the life on the bark: miniature forests of twigs, thin blankets of moss and, as I reach the crown, lichens that decorate the smoother branch-ends in irregular smudges. The lichens are soaked, engorged and shining in the drizzle; shades of cream and mustard that glow surreally in the gloom. They draw my eye but make every step a gamble, the limbs slick and untrustworthy.

I remind myself, though, to stop and take a moment to appreciate this unique perspective. The canopy of Ashtead Common stretches away in every direction in a homely, rain-damp palette. Neighbouring my big oak is a stand of birch, the bare twigs a grey haze tinged with burgundy. Beyond them are groves of younger oaks and ragged thickets of hawthorn, and lively willows that shuffle fresh and green even while stripped of their leaves. The trees are interspersed with big open areas and from here it could be a map, the glades like oceans between continents of trees. A wide clearing is covered by last summer's brown, flattened bracken with a ghostly armada of dead oaks dotted across it, their bleached hulks the result of a fire decades ago.

Walking among the denser areas of the common, it feels like a woodland. But from here, dozens more big old oaks like this one emerge from the younger trees surrounding them. I can easily pick out the biggest, the King Oak, with its enormous globe of a crown prominent above the clearing, its whole architecture visible and the squat trunk supporting half a dozen splayed stems. These old oaks provide the key to understanding the bracken beds and grassy patches and a landscape lost both here, with the development of the younger cohort of trees surrounding them, and to our memory, despite it being more deeply entwined with human history even than coppiced woodland.

This is wood pasture, a form of ancient woodland that is less well known, harder to define and categorise, and, even in conservation, often misunderstood. Wood pasture, though, as Oliver Rackham wrote, 'runs deep in human nature'.[1] For thousands of years, cattle and other livestock grazed among the trees at places like this, so the land provided meat and dairy products as well as wood. Its use by people resulted in complex landscapes where long-lived trees rise over grazed, open areas along with thickets of scrub and young trees, or heathy blocks of gorse; today, wood pasture is often home to our oldest, most ecologically important – and most vulnerable – individual trees, as well as forming an incredibly rich wider ecosystem.

The immense range of wildlife found in the world of wood pasture also helps us understand the pre-agricultural landscape in which all British species evolved, as we'll see in the next chapter, with implications for how we can better manage both woodland and the broader landscape for nature. But my introduction to

Ashtead Common came through my involvement with a project to conserve these remarkable old oaks. They're extraordinary ecosystems in their own right, with their decaying wood supporting a tremendous variety of invertebrate life, as well as highly significant human artefacts. Both their charismatic appearance and their longevity are the product of human use – and the regular attention of previous woodlanders – over hundreds of years, as generations of commoners exercised their ancient rights to take what they needed from the land.

I'm still getting used to this work and have to steel myself to venture out to the tip of each limb; it's rare in the more brutal tree surgery I'm used to, where we lop off whole chunks that pose a hazard, or dismantle entire trees. Pruning the sides of these big crowns is fun, though, if delicate. With my ropes anchored over a solid fork high up I can drop down and lope out along the branches, leaning into my harness and using feet and hands to fight the gentle pendulum effect. There's a compelling theory that early humans learned to stand upright not upon descending from the trees as we evolved from our primate ancestors, but actually to move around them more efficiently, feet on a lower branch and hands grasping those above. Safe beneath a high anchor and moving with something approaching grace along rough oak boughs, I feel the inheritance of these arboreal origins as I stretch out to make my cuts.

Working in the top of the crown is a different story. Here I climb tentatively, making my way up increasingly slender stems and placing each clumsy boot with care. I finally loop a short strop of rope around my perch, hugging the same thin branch that I reach up and cut with my pruning saw, showering myself with shaken drips and a light sprinkling of sawdust. I trim the crown by just a couple of metres at each extremity, enough to reduce its chances of being caught and pulled apart by the wind while maintaining plenty of life-giving foliage. Much like cutting coppice, I grab the twiggy pole as it starts to fall but this time give it a boost to send it out and away, and watch it twist gently to the woodland floor 20 metres below.

The tree I'm perched in is a 'pollard', along with more than 1,000 of its companions here. Like coppiced woodland, pollards were cut regularly throughout history for their crop of useful wood; they were lopped just above head height with the lowest 2 or 3 metres of trunk – called the 'bolling' – springing back to life despite its frequent beheading ('poll' is an old German word for head that found its way to medieval English; 'bolling' comes from the same source – it refers to a round vessel and is a close relation of 'bowl'). Cutting high meant that the lush regrowth was out of reach of the livestock that continued to graze among the trees, creating woody, three-dimensional landscapes that maximised the productivity of both the trees and the ground vegetation.

The onset of the Industrial Revolution marked the beginning of the end of pollard management, as it did that of coppicing. On many wooded commons like this the effect was exacerbated by the Enclosure Movement, which ended common rights and severed the connection between the majority of British people and the land they lived on. The subsequent abandonment of pollarding has seen the trees' multiple poles grow out, in some cases, into the biggest crowns we ever see, like a group of mature trees perched on each individual bolling. This recent, giant growth, combined with the decaying wood in their old bollings, makes them uniquely vulnerable to collapse. Decay, though, is a rare and incredibly valuable ecosystem in its own right; lots of invertebrates rely on dead and decomposing wood for some aspect of their life cycle, including many extremely rare species. With no successor generation of trees developing the same conditions, it's essential that we keep Ashtead's current ancient oaks upright for as long as we can, to sustain their unique value – a value that was maintained, and is now at risk of being lost, as a result of the human history of wood pasture.

*

From the Bronze Age, when humans began to cut trees at scale, until well into the nineteenth century, animals grazed among working trees across swathes of Britain. Wood pasture is both a form of woodland management and an agricultural system, and its effectiveness in providing so many of our ancestors' needs exposes the false distinction between the two that we've come to take for granted over the last couple of centuries. It probably developed as the first farmers cut trees for fodder and noticed that they grew back, and over the first few thousand years of farming in Britain the gradual modification of much of the prior landscape – the wildwood – into managed wood pasture would have been just as important as the thorough clearance of trees in other areas to create fields for cultivating crops; one researcher has suggested that early farmers would have needed an area for 'forest browsing' ten times the size of their cleared fields.[2]

It's thought that, by the time the Romans arrived, around three-quarters of the British landscape had been heavily influenced by people. True to type, they probably separated a greater proportion of the landscape into more intensively managed agricultural fields and coppicewoods, resulting in a lower area of wood pasture as the monocultures provided more efficiently for the needs of a growing population in bigger towns and cities, as well as for sophisticated industries like iron-working – although large areas of shared grazing certainly persisted.

In the less organised centuries that followed, trees recolonised some of the Roman farmland and wood pasture resumed an even more integral place in the lives of many – perhaps most – people. The peasants who formed the majority of the population were only one step removed from earlier tribespeople in living almost exclusively off the countryside around them, and their rights to do so were formalised under the Anglo-Saxon feudal system. The custom remained that landowners wouldn't exploit their holding to the detriment of the communities that relied on it, although this wasn't purely altruistic: the lords of the manor also relied on the labour of the freemen and serfs, not to mention their military service. The most fertile ground close to villages was typically parcelled into fields but the land further out that was too steep, wet or rocky to farm remained a vital shared resource.

With the Norman Conquest, these deep-rooted traditions were further codified into law in England. (The differing stories of land in England, Scotland and Wales confuse the story. I've focussed on England as the home of my featured woods and the only country where I've had practical experience of working them.) William divided much of his new land between himself, the Church and the barons who'd helped him defeat Harold at Hastings. But in many places where the custom was of shared use since 'time out of mind' – beyond memory – the land was held 'in common' and local people were able to carry on using it under closely defined rights that ensured it wasn't overexploited; at the peak of 'commoning' in the couple of centuries that followed, around half of England was common land.[3]

According to the Domesday Book, the comprehensive record of landownership commissioned by William in 1086, woodland cover had fallen to around 15 per cent of England; this figure included some types of wood pasture that were recorded in the survey and, based on those places where the book is particularly detailed, wood pasture could well have been the dominant type of woodland at this point. Understanding the historic extent of wood pasture is incredibly difficult, though, even from a close reading of experts like Oliver Rackham and through conversation with some of his colleagues. Although Rackham asserts that the 15 per cent 'woodland cover' in the Domesday Book included wood pasture, for instance, it's clear that this figure doesn't include all common land, much of which would have also combined grazing with trees, typically pollards. George Peterken states in *Trees and Woodlands* that 'Wooded commons were once everywhere, and remained so well into historical times', and quotes the Welsh historian William Linnard, who confirms that they were 'indispensable to every community'.[4]

Although some landowners did begin to try to exclude commoners, their various rights were protected by Magna Carta in 1215 and the Charter of the Forest two years later. The evocative term 'pannage' is now well known among historically minded conservationists as the right to graze pigs, an important function of oak and beech wood pastures with their periodic mast years, when the trees produce a glut of swine-fattening acorns or beechnuts. The size of a wood or wood pasture was sometimes measured by the number of pigs it could support, although the simple right of pasture for cows and sheep was probably more important. The right of 'turbary' allowed the cutting of sods and peat for fuel, while 'piscary' defined the right to fish from lakes and streams. The rights to bigger timber trees nearly always remained with the landowner but pollarding was covered by the right of 'estovers', which allowed commoners to cut wood for fuel as well as bracken, heather and other vegetation, while in some places there were also specific rights for 'housebote' (wood for building), 'ploughbote' (wood for farm equipment) and 'haybote' (wood for fencing), among others.

It seems likely that a higher proportion of pollard-wood than coppice was used for fuel, and, being so closely integrated with the management of livestock, the poles were also often cut in leaf, dried in bundles and stored as 'tree hay', a nutrient-rich supply of fodder to supplement grass and browse over winter, after which the wood could be burned or processed. In the Lake District, I met a couple of farmers who still cut their ash pollards in the summer. Although they harvest them primarily for firewood and don't dry and store the foliage, the farmers purposely leave the fresh-cut poles in the fields for sheep to strip, having inherited the understanding from their parents that it's good for the health of the animals.

Ashtead Common is, as the name suggests, a relic of this ancient system of commoning – it probably remained as shared, rough grazing because its clay soils lie sodden all winter, as I can well attest, so it wasn't much good for more intensive farming ('ash' could be derived here from the Old English 'eis', for a spring or water, as it seems unlikely ash trees were ever dominant on the common). The pollards have been aged to around four or five hundred years old, making it a remnant medieval landscape that also easily qualifies as ancient woodland; it's clearly been wooded, after a fashion, since these trees were first beheaded as saplings.

As with most land in Britain, it's difficult to know for sure how tree cover developed here prior to the seventeenth century and the only clues we have are beneath the ground. In one corner of the common lie the remains of a Roman villa and tile factory, with the implication that the area might have been managed as coppice to provide fuel for the tile kilns – as well as for the villa's under-floor heating system, which I saw perfectly preserved during an archaeological dig one summer. Just beneath the King Oak, conversely, there's a triangular earthwork that dates from before the Romans and might have been used to enclose livestock, hinting at a more open landscape in the Iron Age.

It's not unusual, on delving deeper into the history of an ancient woodland, to find such evidence for radical changes in land use before 1600. This means that the classification of 'ancient woodland', while useful, is never simple. In some respects, what matters in ecological terms is primarily that these ancient places haven't been ploughed for many centuries, or, perhaps even more importantly, sprayed with pesticides or artificial fertilisers; as long as the soil has gone undisturbed for such a long time, much of its life will persist despite the tree cover coming and going.

On Natural England's Magic Map website, Ashtead's outline surrounds the vertical green stripes indicating ancient woodland, as do many other historic wood pasture sites. Lots of ancient wood pasture was included in the original ancient woodland inventory in the 1980s, although without any distinction being made between woodland and wood pasture. Even more sites have been incorporated into the recent review, and where possible they have been recorded separately as ancient wood pasture; but even beyond those places categorised by their history since 1600, much more ancient woodland is likely to have been more open before its conversion to coppice over the course of the previous millennium, the landscape in a process of perpetual change in response to the needs of people.

Once you learn to read a little of the language of the woods, Ashtead's pollards alone provide a glimpse of the way local people used the common for hundreds of years to provide both wood and grazing; their current form, however, articulates even more powerfully the decline of wood pasture over the past couple of centuries, illustrating how these misunderstood landscapes and their special trees came to look as they do today. The size of the big stems on the old trees – typically fatter than post-boxes – suggests that some of the poles themselves have been growing for well over a hundred years, dating the point at which the much older bollings were last cut to some time in the nineteenth century. The abandonment of the pollards could easily relate directly to the construction of the railway to Ashtead in 1859, which brought cheap coal to all and made their harvest for firewood redundant. Cattle might have grazed the common a little later but the age of the younger generation of trees suggests saplings weren't nibbled for long, as Ashtead evolved into a commuter town rather than a rural village. Historic maps I found online seem to confirm this: the area of the common marked with the scattered tree symbols of wood pasture declines even between maps made in 1871 and 1896, and is replaced with the dense stamps of woodland.

As its direct utility for local people ended, Ashtead Common survived because it provided something more intangible, although no less important. The removal of common rights had begun before Magna Carta but the Enclosure Movement of the eighteenth and nineteenth centuries eventually sounded the death knell for wood pasture as a major use of land, along with the movement's driving forces of industrialisation and the increasing market economy. Between 1760 and 1870 nearly 3 million hectares of common land – around a sixth of England – was enclosed at 4,000 individual sites, each of them requiring an Act of Parliament that removed the rights of the commoners and gave the owner the right to use the land as they pleased.[5] Typically this meant 'improving' the land by clearing, ploughing and fertilising it to focus more profitably on commercial agriculture, but some commons were converted to woodland under coppice management or, as time went on, to forestry plantations.

Wooded commons were described by landowners as 'waste' or 'scrub', in contrast to the modern 'scientific' systems they could become, while commoners were presented as either the starving poor or idle layabouts – both in need of work. It's probably obvious that most of those who lobbied for Enclosure were the very people who stood to gain as they came to exert complete control over the land and to benefit from the new systems of land use. What they interpreted as poverty and idleness was, in many cases, simply that commoners didn't need to rely on full-time paid employment – still a far-from-universal concept – because of their access to common rights, local systems of barter and ancient traditions of cooperative labour.

Whatever the reality of the lives of commoners – and it is, of course, important not to romanticise their essentially medieval existence – the landowners were frank about wanting to 'help' them out of their supposed poverty or sloth and into paid work, either as labourers on the newly enclosed land or in rapidly developing urban industries; that is to say, in the service of the landowners. Enclosure resulted in the disenfranchisement of millions of ordinary rural people and the concentration of land, the ultimate resource, in the hands of a powerful few, as well as contributing to the simplification of the complex ancient landscape into open fields and dense woodland. Instead of providing a wide range of products for local people under wood pasture, huge areas of common were converted to produce a much smaller variety of goods that were exported for sale elsewhere.

As the Enclosure Movement progressed into the mid-nineteenth century, some landowners around London realised that they could most profitably benefit from the removal of common rights by selling their land for built development as the city sprawled into the surrounding countryside. It was the impact of this trend on the middle classes that brought the era of enclosure to an end as people began to recognise the loss of green space on which they relied for recreation, resulting in the formation of the Commons Preservation Society in 1865, which in turn paved the way for the National Trust and for the City of London Corporation's step into the care of the countryside.

Early in its history, following the enclosure of Hainault Forest in north-east London and its conversion to low-quality agricultural land, the society successfully campaigned against the enclosure of the neighbouring Epping Forest, an important historic wood pasture that remains home to Britain's most dense concentration of pollarded trees. In 1878 the Corporation agreed to take on the ownership of Epping and to manage the land under an Act of Parliament that required it to 'at all times keep Epping Forest unenclosed and unbuilt on as an open space for the recreation and enjoyment of the people'. This was, essentially, a philanthropic gesture by the wealthy bankers, industrialists and merchants who ran the institution, and the Corporation went on to agree to care for other commons and open spaces that were typically saved from Enclosure and development by public campaigns, like Hampstead Heath and – somewhat less well known – Spring Park.

In 1875, a couple of years before Epping was formally saved, the same movement had become influential enough that parliament passed an act stating that enclosure would only be permitted where there was demonstrable public benefit. The remaining commons were preserved as open spaces and many that we know today, particularly lowland wooded commons, survive due to the success of the Commons Preservation Society and similar organisations.

Ashtead Common remained in private hands but unenclosed during this period before coming into the ownership of Vernon Willey, 2nd Baron Barnby. He had no real option to enclose the common but also encouraged its use by local people, and after the death of Lady Barnby in 1988 their trustees, acting on the couple's wishes, agreed the sale of the common to the City of London for management in accordance with the Open Spaces Act. The common today remains a common in law as well as in name, which – Open Spaces Act notwithstanding – would require the agreement of parliament to change its use or even to erect a permanent fence. A couple of local people still have common rights of estovers linked to their properties, which the rangers satisfy by dropping trailer-loads of firewood created as a by-product of conservation work.

Wooded commons aren't the only form of ancient wood pasture that played a role in defining the landscape of Britain. In parallel with the designation of common land after the Norman Conquest, the wider story of wood pasture became much more convoluted. Large areas – including commons – were 'emparked' with fences or walls to enclose fallow deer, imported from the Middle East. Other areas were designated as forests and chases where native game was legally protected for hunting by the royals and aristocracy, as well as simply to provide for their lavish feasts.

The word 'forest' only later came to mean a large area of woodland, typically a plantation, and here defined an administrative area subject to particular legislation. Forest law has passed into legend for its severity, with punishments for transgressions including the gouging of eyes and severing of hands. Oliver Rackham points out that these brutal penalties were rarely enforced and that the benefit to the allies of royalty in being bestowed a forest or chase was that its laws provided income, as well as ownership providing status; the rights of local people to harvest daily essentials were maintained in many places and the fines levied for misdemeanours such as poaching, stealing wood or illegally grazing animals were typically in line with their value.[6] The threat of gruesome punishments served as an incentive to pay up.

Pollarding was stopped in some deer parks and forests as medieval landowners began to focus on land uses from which they could profit, with some areas within each park, for instance, typically being converted to timber and coppice. But broadly speaking these layers of aristocratic protection slowed the clearance of wood pasture, while the countryside beyond began to be more rapidly cleared and 'improved' for agriculture during the remainder of the Middle Ages and into the modern era. Some of the best-preserved wood pasture remaining in Britain today, like the New Forest, is the result of its history as both a royal forest and a common, and lots of local people still exercise their rights to graze livestock.

The incorporation of many medieval deer parks – as well as other areas of countryside – into ornamental parks around stately homes also played a vital role in preserving wood pasture. Almost uniquely to Britain, the influence on the design of these landscapes of the Picturesque movement, with its focus on inspiring awe and evoking a sense of antiquity, helped preserve elements of the prior wood pasture landscapes that we've now come to realise are particularly important, chief among them being their old trees. Prominent landscape designers of the eighteenth century like Humphry Repton and Capability Brown were known for incorporating big existing trees into their parks from the working landscapes that pre-dated them, valuing the trees for their aesthetic appeal and their ability to lend a sense of permanence; this at a time when, in equivalent projects on the Continent, previous landscapes were typically razed to make way for more formal, entirely planted creations (a trend reflected in the many avenues planted in British landscape parks, even alongside more naturalistic elements). Historic parklands, like those at Dunham Massey near Manchester, where I worked, Windsor Great Park and Wimpole in Cambridgeshire (which developed under the influence of both Brown and Repton) remain today some of our best-preserved and most recognisable wood pasture landscapes.

It's also easy to overlook the fact that, throughout most of human history, even the farmland outside what we'd now call wood pasture or woodland would have been a much more varied landscape of open spaces among trees. Until the post-war industrialisation of agriculture, most fields were small and would have been surrounded by hedges and trees even in formal agricultural landscapes, and there would have been many more trees in the fields – as we see in the occasional valleys in the Lake District or Eryri that have escaped the worst effects of intensification. The whole landscape remained a more intricate patchwork than we're accustomed to now, with lots more trees and most pieces of land used to provide a range of crops and products.

The braided history of the various strands of wood pasture – outside the ornamental landscape parks – came back together, in the broadest terms, from around the seventeenth century, following the slow erosion of most hunting forests and culminating with the main period of Enclosure a couple of hundred years later. Regardless of whether it had been previously designated a common, chase or forest, the multi-layered, multifunctional wood pasture that had played such an integral role in the British landscape began to be compartmentalised increasingly rapidly, in most places, into the binary landscape we know today. Fields were made more open and transformed from rough pasture to arable using developments in agricultural technology, and other areas of wood pasture were converted to coppice or to the newly developed plantation forestry: an Act of Parliament in 1660 had paved the way for the latter by overriding rights to other land uses in the Forest of Dean to make way for the oak plantations that were famously intended to provide for the navy. Even where more marginal land could only sustain grazing, trees were cleared to allow mechanical haymaking, the 'improvement' of the grass-crop by fertilisation and sowing new varieties, or simply because newer, more productive breeds of livestock were less hardy and couldn't cope with bramble and scrub.

The wood pasture landscapes that survived into the twentieth century were typically those where a relatively self-sufficient way of life continued despite enclosure and the influence of the market economy, like on the upland farms of the Lake District, or where they'd been incorporated into the landscape parks of the aristocracy; or, as we've seen, where commons were preserved on behalf of a wider range of people for the new common goods of recreation and open space, like at Ashtead.

As I look down from my wobbly tree-top stance on the continents of trees and their intervening oceans of bracken, grass and bramble, it's mind-boggling to think of the changes to the common's landscape over the course of centuries and more. Our current moment in history is the first time in at least 500 years – and almost certainly much longer – that Ashtead has witnessed a canopy as tall and closed even as this. The site's history since the end of the nineteenth century illustrates the story of many wooded commons that were preserved for people to enjoy but whose underwood and grazing became redundant. Unchecked by livestock, wind-borne birch and willow seed – as well as the seeds of hawthorn and blackthorn, deposited by birds in their droppings – began to colonise the open areas, followed in short order by a younger generation of oaks that have matured to turn parts of the common into something more recognisable as woodland.

The pollards too, have bloomed in their abandonment, for better and worse. Prior to the end of pollarding they were each just 2- or 3-metre stubs of trunk scattered across the open landscape, their crowns limited to another few bushy metres of arm-thick poles between each harvest. The light weight of the young regrowth was all the swollen bollings were required to support throughout their centuries of working life, so, even as they began to hollow with age, their burden was relieved every decade or two and they could persist almost indefinitely. (The insignificant load of the crown of an 'in cycle' pollard, combined with the low profile it presents to the wind, means that pollards can probably outlast the natural lifespan of 'maiden' trees of the same species; some of our oldest remaining ancient oaks – which may be 1,000 years old – are likely to have been pollarded for at least some of their lives.)

Neglected, however, the poles that sprouted after the last cut have grown into huge, tightly packed groves perched atop the decrepit bollings, each one a mature tree in its own right. They arc and spread from the bollings, two or three, sometimes five or six, some shooting straight up while others spread like a candelabra, and all of them branching and forking into typically complex oak-tops. Some of the poles on Ashtead's pollards are old and rotund enough to have developed their own holes and cavities, while the combined crowns of the biggest overstood pollards have become more voluminous than even the most spreading uncut oak, making them some of the most beautiful and awe-inspiring trees in our landscapes.

The bolling of the oak I'm in is particularly bulbous and haggard, and there are just two thick poles arcing away from the collapsing base. The remains of another lie beneath a dun heap of dead bracken, its butt end flared and torn, and I think at least another one must have failed longer ago: I made a close inspection of what's left before committing to the climb. Gargoyles of exposed heartwood lend a sculptural quality to the bolling, while dry red rot is visible through gaping holes where the old poles have sheared away. There are crumbling, cubic columns, the colour of milk chocolate or lightly roasted coffee, where the microscopic hyphae of fungi like chicken of the woods and beefsteak have wormed their way among the very grain of the wood to consume it over decades and centuries. The tree persists even as its core hollows and collapses, the new growth at its outer edge keeping pace with the slow degradation of the oldest wood.

The wood and fungi become more than the sum of their parts in ecological terms, combining to become almost an extension of the woodland soil with its webs of mycelium and constant cycles of growth and decomposition, and the decaying mass is alive with invertebrates that rely on every minute aspect of the process, each with its own demanding needs. Decaying wood is integral to at least part of the life cycle of more than 2,000 'saproxylic' invertebrates (literally 'of rotten wood'), 7 per cent of Britain's entire invertebrate fauna; David Attenborough said of trees like this that 'there is little else on earth that plays host to such a rich community of life in a single living organism'.

Some species use the obvious spots, under loose bark or beneath fallen trunks, but the requirements of many other species become almost absurdly specific. There are hundreds of beetles that rely on the work of fungi to break down tough, resistant wood before they're able to burrow galleries – branched tunnels – in which to lay their eggs, and each of these has evolved to rely on particular tree species, parts of the tree, types of fungi and stages of decay, from the brittle brown decay caused by beefsteak fungi in oak to the soft, wet 'white rot' that *Ganoderma* causes in beech.

Some beetles even carry fungal spores and inoculate the wood in which they lay their eggs, while others make their homes in the fruiting bodies of the fungi themselves. There are beetles that use the galleries left by previous species, and also a whole range of even more rare predators and scavengers forming a saproxylic food web in each trunk. Cobweb beetles feed on the scraps left by spiders that live beneath loose bark, while there's a beetle known only from Windsor Great Park whose larvae prey on the hairy fungus beetle, which itself feeds on the thick, shammy-like mycelium of chicken of the woods fungus as it works its way through the heartwood of oak. The life within each tree is almost inconceivable, the infinite variety and high drama of the Serengeti or the Amazon playing out in miniature. As well as the beetles there are flies that rely on water pockets in which to lay their eggs, pseudoscorpions scurrying beneath dead bark and two species of tree ant that live in decayed heartwood and whose nests provide habitat for other specialists.[7]

We now associate some of them with human structures, but lots of less obscure species like swifts, honey bees and most British bats evolved to use holes in old trees, among many others. The richness of the ecosystem increases with age, and the juxtaposition of the trees' continuing life and growth with dysfunction and decay make them some of the most significant keystone organisms in Britain.

Like our oldest and most important woods, we typically call them 'ancient trees' to reflect their value. Oliver Rackham famously wrote that 'ten thousand oaks of one hundred years old are not a substitute for one five-hundred-year-old oak', emphasising the irreplaceable ecological value of all those species that rely on decaying wood.[8] There are more ancient trees in Britain than in the rest of Europe combined, making them one of our unique responsibilities towards global conservation. (On the Continent, centuries of warfare saw desperation periodically overtake the sustainable management of pollards in many places, and the bollings chopped up for firewood. The English tradition of the eldest son inheriting an estate also meant many wood pastures stayed intact where in other countries they were divided among siblings, while those different trends in designed landscapes also played a part.)

Here at Ashtead, the size and sail of the overstood poles on their shaky foundation means the pollards are at increasing risk of collapse as the bollings decay while their crowns increase in size. Some are blown apart by November gales, but others seem to simply give up and are discovered root-heaved or branch-torn on still summer mornings. Freshly turned earth beneath naked, upheaved roots, bright oversized splinters and livid cracks wet with sap indicate a moment of unwitnessed violence and the gradual degradation of the common's value for wildlife. The rangers monitor the failures and know we're losing at least 1 per cent of Ashtead's ancient trees each year, so there'll be few left by the turn of the next century; it's an impossibly distant future for us, but nothing to an oak or its internal ecosystem. There aren't enough trees growing alongside them that will reach old age soon enough to develop the decay features on which the rare beetles and other invertebrates rely, so we need to keep the oldest oaks going for as long as we can in the hope that there'll be more replacement habitat before they're gone.

Our main tactic to reduce the chance of failure is simply to prune them, each cut an act of care. In the 1990s, when conservationists first realised both the significance of ancient pollards and their inherent instability, some trees elsewhere were completely repollarded, their poles lopped back to the bolling. Unfortunately, the bark of canopy species like oak and beech typically becomes so thick and fibrous with age that dormant buds weren't able to burst from the tops of the bollings, so many of the repollarded trees died – a hard lesson learned. We know now that we need to treat the pollards with more sensitivity.

For the past few years I've joined arborists Richard and Jamie, who spend weeks at Ashtead each winter and are some of the most experienced climbers in the country when it comes to the care of our oldest trees; in all the work I do here, I'm an apprentice to them. Richard is older, quiet and gentle. He wears overalls and a flat cap when he's not in his helmet and his role is typically to prune the sides of the trees from a 15-metre-tall cherry picker, an articulated platform that folds up to crawl around the common on caterpillar tracks. When he does need to climb, though, he's deceptively nimble; I somehow rarely catch him on the move but every time I look up he's somewhere new, positioned solidly for his work. Jamie, in contrast, is a ball of virtuosic energy who climbs to the top of the crown to prune everywhere Richard's platform can't reach. He'd greeted me on my first morning with a one-two shadow-box that had me flinching before he swung a friendly arm around my neck: 'Charlie Barley is it – ready for some *hardcore ancient tree action?!*'

This job is, they told me, the highlight of their year and what Richard calls 'the cream' in comparison to the type of routine tree surgery we're all usually involved with; it's off in the wood away from roads and people and, not only are we working on some of the biggest and most impressive trees in the country, we're doing it for their benefit rather than to remove a risk or nuisance. Richard and Jamie explained that we should cut off just a couple of metres at the end of each branch, an unusually light reduction for trees of this size. Even this small intervention can, they said, reduce the effect of the wind on the crown to a disproportionate degree, but it leaves more than enough foliage on the tree to keep it alive and can also help encourage the development of a more sustainable new crown lower down.

It was intimidating when I started to work alongside the two of them, knowing I have to get right out to the end of every branch of the largest trees I've ever climbed and then make good decisions about where to cut, both for the sake of the tree and under Jamie's passionate, perfectionist eye – and all while trying not to slow the steady progress of the team's annual campaign across the common. The biggest trees, like the King Oak with its half a dozen wide-crowned poles, take a couple of days of solid climbing even with Richard working the sides and Jamie in the top, and I spend entire days in some of the pollards I prune, becoming intimately familiar with their geography. There's no language for the different parts of their crowns as I mentally divide them into sections, no name for the anchor points on which I come to rely, but the great scaffolds become intimately familiar for the duration of my time in them, at my best; albeit sometimes with an element of dread at worst, a big limb put off until later for the awkward, precarious stretch across to it, or the discomforting sense of exposure as I straddle a horizontal stem for want of a better stance.

Despite his high standards and relentless work ethic, Jamie's always supportive and helps me realise I don't need to emulate his nimble style. My confidence increases as I learn to approach each oak strategically, trusting my ability to spot anchors that will help me ascend most efficiently and get around the crown without having to relocate them, and honing my ability to launch a 'handbag' of tightly coiled rope over the right branch high above, tugging it in just such a way that it unfurls back toward me. Richard shows me tricks to send loops out to catch the hanging end if it's out of reach, and as time goes on I begin to feel that, while I could never say I'm fast, I am at least a help rather than a hindrance. 'You get up there and you get on with it,' Jamie reassures me, clapping a hand on my shoulder, 'that's three more trees helped today!'

When I've visited every extremity of its crown and finally finished pruning my oak, I can barely remember where I started. The first, cold cuts this morning seem a lifetime away and, heated by exertion, my wet shirt and aching forearms feel weirdly comfortable. I sit back in my harness and shout to the rangers clearing brash below to check I've not left any daft tufts sticking out; they have a much better view of the whole crown than I do. Grabbing the locking knots that secure me to the ropes subtly loosens them and I descend at the speed just before friction burns my palms, stopping once to push off from the trunk and swinging out to grab a cut branch that's tangled on its way down.

My last stop, by habit, is in the big union where the giant poles spread away from the old bolling. On this tree I can't stand in the tight gap between the two remaining stems, but find footholds either side of the wounds left by the sections that have torn out. Someone stood in this spot a dozen or more times throughout the oak's two- or three-hundred-year working life, swinging an axe or billhook to fell the poles. While our aim in pruning it now is to preserve its extraordinary habitat, this tree is a human relic, created by the actions of people as they exercised their timeless rights to take the wood they needed.

Each of those woodlanders might have cut this pollard three or four times during their life, no more. Perhaps they were related and generations of a local family tended to it, passing their knowledge of its idiosyncrasies down the line. Perhaps it was a reliable tree that sprang back even if left a little longer for thicker poles, or maybe it needed cutting a little early to ensure good regrowth. I feel sure they would have known. I wonder where the wood went, and whether any of it persists as a lintel or joist in a building nearby, or in a tool handle or plough frame preserved in a barn or museum. Whoever cut it would probably have used the wood themselves, possessed of an unimaginable depth of knowledge to convert the fresh poles into the things they needed. We talk among the crew about all this, and about what the common must have looked like when the tree was last cut, and, inevitably, about what the commoners would have eaten for their lunch. We wonder what they would have thought about our work – and our equipment – today. We assume they'd be completely mystified that we're working for something so abstract as the trees themselves and the life they support, rather than for the essentials we need to survive in a more immediate sense.

It's daunting, at first, to prune these amazing beings that instinct tells me should be preserved untouched. I hope our work is successful in slowing their rate of failure while sustaining the trees' health; we'll have to wait years to properly analyse any changes and probably won't know for decades whether our graft is worthwhile. But not only do I want to look after this tree for its inherent value and as one of Britain's most unusual and vulnerable ecosystems, in writing the most recent chapter of its story with my saw I want to honour the woodlanders who wrote, with their tools, what came before. Their work and lives went unrecorded, too ordinary and mundane to be worth the time and attention of those with the power of the written word. These trees are all they left.

And as well as the commoners' careful influence over hundreds of years, the pollards also speak to me of the injustice of Enclosure; the characterful forms caused by their abandonment are a living reminder of the subsequent disconnection between people and the natural world. Until that period of severance we were all woodlanders, with perhaps the majority of people throughout history having spent their lives cutting pollards like these on common land like this.

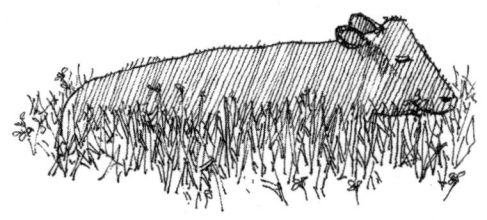

CHAPTER FIVE

A Glimpse of the Wildwood

The landscape before farming

THE AIR ERUPTS into a sparkling cloud as the low sun is magnified by drifts of willow seed, caught by an imperceptible breath of wind and now floating dense around me. I squint into the glare, half blinded but hypnotised by the shifting light. Suddenly there's another element, as I realise the scene is soundtracked by what seems like dozens of willow warblers, their soothing, descending whistles overlapping from every direction. Although they're never particularly unusual to hear as individuals in the woods I help look after across South London, I'm still amazed to find myself amidst this near-flock of these subtle little birds. It is both a reminder of what we've lost, with the warblers having been in decline for decades, and a powerful illustration of the value for wildlife in wood pasture landscapes like this at Ashtead Common.

I've been checking the cattle hidden among the bramble and scrub, a twice-daily task that the rangers take turns at on summer evenings. Downhill from the main concentration of ancient oaks, parts of the common were cleared to grow potatoes during the Second World War. Abandoned once more since then, they've slowly developed into unkempt patches of hawthorn, blackthorn and willow, a homely landscape that somehow brings an inherent sense of peace as I wander through it on patrols that hardly feel like work. The small herd of Sussex red cattle have been introduced to break up the scrub and drive a constant process of change in the intricate combination of thorny growth and open space. While it's easy to assume that working commons were relatively simple landscapes of pollards over wide areas of grazing, we know now that wood pasture is at its most valuable for wildlife when, like this, it's a more complex blend of open habitats, scrub species and trees of all ages; at its best, in fact, it is perhaps the richest type of British woodland.

Over the last twenty years or so, conservationists have combined their increasingly detailed knowledge of the needs of a wide range of species with the experience of seeing new wood pasture habitats develop in rewilding projects. We've come to realise that the reason wood pasture is so important for wildlife today is that it probably looks – and behaves – something like the pre-agricultural wildwood in which all species evolved. As well as emphasising the importance of historic wood pasture and helping us understand the importance of rewilding, this knowledge also helps us better appreciate the needs of our threatened woodland wildlife and sheds light on why coppicing was inadvertently so important for nature. It can inform the future management of both ancient woods and wood pasture, as well as the wider British landscape.

The cattle aren't alone in their work at Ashtead: humans give them a helping hand. As well as Richard and Jamie working in the pollards each winter there's a little crew of contract tree-fellers working on the ground. They're led by Jim, who wears a Rocket from the Crypt patch on his jacket and whose friendly nature belies his punk-rock work ethic. On the best days a dozen of us rendezvous at the common's top gate just as it's getting light, pulling off the busy road between the M25 and central London into the gloom of the trees. We park our trucks to avoid them sinking in the wet clay soil and disappear into the common on foot like a medieval caravan, a line of orange-helmeted woodlanders pushing wheelbarrows or simply laden with rope bags, climbing kit, chainsaws and fuel, and with the tracked chipper and cherry picker trundling along like ungainly beasts.

The tree-fellers typically operate just far enough from the climbers that we don't get in one another's way. Before most of the pollards can be pruned, Jim and his team open them up by removing a ring of surrounding trees. The vigorous young upstarts that have grown since grazing stopped in the nineteenth century can overtop the pollards, shading them out as well as competing for moisture in the soil, and the lower branches on some of the old trees have even died due to the lack of light.

Felling younger trees around the pollards has the additional benefit of opening up the common as a whole and beginning to restore its health as wood pasture; in the years that follow, I'll realise that this simple, uncelebrated work is just as important as the dramatic pruning of the oaks in regenerating this special ecosystem. As the historic landscape begins to emerge around the oaks thanks to the work of Jim and co., the cattle will be encouraged to graze up there, too, bringing the whole place into a system that reflects its working past but will also be significantly better for nature.

Lying at the heart of the great value wood pasture holds for nature is the fact that it isn't actually a single habitat, so we find species associated with a wide range of individual environments all thriving in close quarters: the flowers and butterflies of unimproved, species-rich grassland or heathland; birds zipping among hawthorns and other scrub; and woodland wildflowers beneath groves of mature – or even ancient – trees, with the myriad life they support. Wetlands, streams and rivers all bring their own unique ecosystems and species to the mix.

But we also find something more magical happening when open habitats, scrub and trees – along with watery places – develop together, something that's been lost from much of our compartmentalised countryside with its distinct boundaries between open fields and shady woods; and particularly with the post-war inclination for tidiness, and the subsequent eradication of scrub. The rare alchemy lies in the interactions *between* habitats that we've come to think of in isolation and, where wood pasture landscapes persist or are undergoing successful restoration, we find a surprising number of species thriving – including some that have been mysteriously disappearing everywhere else.

Although not rare, the green woodpecker is emblematic of wood pasture as it feeds on yellow ants from their mounds in unimproved grassland before speeding, with its unmistakeable yaffling cry, back to mature trees nearby. Many of the species we think of as garden birds evolved to need a similar combination of cover and open feeding grounds, flitting from dense scrub to delve for worms and other invertebrates beneath grassland – the familiar robin and blackbird have adapted to use the cover of garden hedges between their stints feeding on lawns. (Adding a bit of woody structure by planting shrub species or trees can easily increase the wildlife value of gardens dominated by grass, which should never be treated with chemicals that affect the soil fauna.)

Typically in greater trouble today are the birds most often associated with woodland. The willow warblers I heard are among a couple of dozen species, also including the nightingale and dunnock, whose decline has been attributed to the end of coppicing in ancient woods. But rather than really being woodland birds, they actually rely primarily on scrubby, fresh growth, like the willows and thorn on the lower common – which, in turn, need bare, disturbed ground in which to grow and aren't strictly woodland species of tree.

While some of the nightingale's last remnant populations were found in the young regrowth of the few woods still in active coppice management, experiments in restoring farmland to something more natural have better illustrated its actual habitat requirements. One of our most endangered birds, iconic for its song, its needs are particular: it only lives in dense growth that's roughly five to eight years old, thick enough to provide cover but before it matures so far that the understorey opens up. Nightingales therefore require a constant cycle of new shrubby growth developing at every site where they breed. Those few remnant woods still in traditional management simply provided a 'good enough' proxy for their preferred home in which the birds could cling on – sometimes called 'refugee habitat'. The explosion of nightingales at the Knepp Estate in West Sussex, though, has shown how the creation and restoration of wood pasture landscapes suits them even better, and could provide a ray of hope for the bird's future.

Knepp has become 'the Graceland of UK conservation', as Tiffany Francis-Baker writes in *The Bridleway*, 'a beacon for anyone weighed down by the burden of ecological crisis and who desperately wants to find a different way forward'.[1] I first visited the now-famous 'wilding' project with Jamie and his mate Ted Green, one of the founders of the Ancient Tree Forum (ATF), a small charity dedicated to the conservation of our oldest trees that I've been involved with since my time at Dunham Massey. Ted played a key role in inspiring the project, and was now busy elsewhere on the estate, pollarding trees in a young plantation of broadleaves to make tree hay. Soon afterwards, I worked with ATF colleagues to help organise a conference where more than a hundred of us spent a day at Knepp and a day at Ashtead; the naturally developing landscape on the former arable farm opened my eyes to just how valuable for nature places like Ashtead can be if we get their restoration right.

While Knepp has become famous as a new 'wild' landscape,

what is perhaps discussed less often – outside the world of conservation, at least – is that the result of Charlie Burrell and Isabella Tree's decision to allow natural processes to drive landscape change is that a young wood pasture has developed, illuminating the true ecological niches of species that we previously associated with woodland or farmland. Nightingales have become one of its headline success stories: they're resurgent in the big areas of thorny scrub. There's always a new thicket to meet their needs as their previous homes mature, and the bird's numbers have boomed from seven 'territories' in 1999 – already a minor stronghold – to forty in 2021 (and more since then, I've been told).[2] The turtle dove is the fastest-declining bird in Britain but is also experiencing a revival at Knepp due to the combination of scrub to nest in, big dead trees that provide important perches from which the male birds sing, and the pig-rooted earth at the edge of wetland areas; plants that thrive in the fresh-turned mud are an important source of food.

The purple emperor butterfly, meanwhile, typically associated with big old oaks that the adults display and mate around, also needs large groves of sallow – a type of willow – where the females are choosy about finding a particular leaf-shape on which to lay their eggs. Another big draw at Knepp, purple emperors are also hanging on at Ashtead thanks to the combination of the ancient oaks with those thickets of willow further down the slope. I had a memorable run-in with two huge, iridescent butterflies gliding down onto the path ahead of me with great grace and spectacle to feast on a pile of dog mess, from which they extract minerals that are lacking in their diet; the males will 'puddle' on faeces as well as dead animals or more savoury pools of water and mud, but they need to fly down to the woodland floor in sunlight, which adds another requirement to our understanding of their needs. Not only have they evolved to need both oaks and sallow, purple emperors also need open areas in which they can descend to the ground.[3] Wood pasture, in other words.

*

The fact that Knepp is a flagship site of the 'rewilding' movement gives the final clue to why the best modern wood pasture landscapes are so rich in nature: the habitat that's developing there probably replicates important aspects of the more natural pre-agricultural landscape, Rackham's wildwood, in which Britain's native species evolved. A better appreciation of the likely workings of the wildwood can help us more fully understand all of their needs.

Until just a couple of decades ago the consensus among ecologists was that the wildwood was a solid canopy of trees: the popular myth went that a squirrel could have scampered from coast to coast without ever needing to touch the ground. This belief was inspired by observations of contemporary British woods that, ungrazed and unmanaged, became tall, dense and dark, as well as by the fact that any neglected field will develop into scrub, then woodland. Ecologists typically believed that 'succession' – this natural process of scrub and trees colonising open ground – was a linear process, with woodland inevitably developing as a 'climax community' at its end, and assumed that the pre-agricultural landscape would have reached this point. Small open areas might have occurred where trees failed through age or by being blown over, according to this theory, but essentially the whole of Britain was thought to be densely wooded, and mainly with oak – which, again, was the result of observation of the composition of existing woods.

This assumption, however, didn't allow for the fact that the wildwood was home to big herbivores and other ecosystem-defining keystone species that created the conditions for many other species to evolve and survive. Some of those keystone species are familiar today, like red deer, wild boar and beavers, and some have been lost, like elk and the aurochs, the wild ancestor of the cow. Both were hunted to extinction in Britain around 3,000 years ago – following the loss of even bigger and more impactful predecessors, as we'll see.

In the late 1990s Frans Vera, a Dutch ecologist, published

a thesis articulating a new vision of the wildwood that took into account the inevitable influence of those animals on their landscape. He was inspired by his observations at the Oostvaardersplassen, a polder, or area of land reclaimed from the sea, that was left to develop as a nature reserve without any human intervention. He saw the impact of geese in maintaining open areas before old breeds of cattle, ponies and red deer were introduced to stand in for wild herbivores. The result was a wood pasture landscape that quickly boomed with wildlife and, although Oostvaardersplassen is just one example – and an imperfect, although fascinating, one at that – the theory that Vera presented, with a wide range of evidence and examples, has been incredibly influential.[4] Insofar as few experts would now suggest that the wildwood was a dense, dark place dominated by trees, 'the Vera Hypothesis' has been broadly accepted – although fierce academic debate continues to rage about the exact balance between woody vegetation and open habitats and their changing state over millennia as studies of prehistoric pollen cores, snail shells, beetle carapaces and other long-preserved remains continue to build a sometimes contradictory picture of this lost world. (I feel compelled to add that in one of my favourite books on ancient woodland, *Woodland Heritage*, published in 1990, Peter Marren describes a very similar version of the wildwood; during my career, everyone has referred to Vera, and his work has undoubtedly had a big impact, but some of these ideas were clearly coming together earlier.)

Vera proposed that the key to the wildwood was in its dynamism, the landscape in a constant process of radical flux under the influence of grazing and browsing animals, with its structure an ever-changing mix of open habitats, scrub and groves of trees. When trees collapsed, due to storms or decay, hungry herbivores would keep those areas open for a while by nibbling any saplings along with the lusher vegetation. If something caused a change in that grazing pressure – disease among the animals, perhaps, or the influence of predators – scrub developed; and within the thorny scrub, protected from any resumption of grazing, new canopy trees became established, beginning the cycle again. Also in the mix were the strong influences of terrain, altitude and soil type, not to mention massive wetlands and bogs, and untamed rivers and deltas made even more vibrant by the work of beavers. Ecologists typically now describe this effect – and its application to modern conservation – as being like a 'mosaic' of habitats, but Rina Quinlan, a rewilding specialist involved with Knepp, pointed out to me that it should more accurately be described as a kaleidoscope, which recognises the central importance of continual change – and has a nice implication of the potential spectacle.

From studies of more natural landscapes elsewhere in Europe and beyond, it's clear that some of these changes happened at a relatively small scale, creating intricate local combinations of what Sophie Yeo describes in *Nature's Ghosts* as 'subtleties of light and shade, depth and edges, youth and rot',[5] while some would have happened at the scale of landscapes or continents as the result of changes in climate or powerful storms. They would all have happened at different spatial and time scales; even in any one place the changes might sometimes have been big, fast and violent, and at other times small or glacially paced.

And it went on for longer than we can begin to grasp. While we often focus on the British landscape that has developed since the end of the most recent Ice Age, other ecologists have pointed out that the influence of earlier megafauna would have been even more significant in defining the processes of their habitats, creating complexity and dynamism at enormous scale during previous inter-glacial periods stretching hundreds of thousands of years and more. Yeo writes that creatures like mammoths, woolly rhinoceros and cave bears 'acted as heavy machinery: a convoy of bulldozers, ploughs and diggers that shaped the ecosystems where they lived'. They would have stripped bark, scraped wallows the size of football pitches, transferred plant seeds and nutrients long distances, and smashed over even the biggest trees – which evolved to coppice in response, millions of years ago. And it was in this landscape, roamed too by straight-tusked elephants and sabre-toothed tigers, that all British species evolved, not in the few thousand years between the last glacial period and the Bronze Age, the period of wildwood referred to most often. It's thought that the human contribution to the extinction of animals like this is probably the biggest influence we've ever exerted on the landscape and the way its ecosystems work, dwarfing even the development of agriculture.

Ultimately, all of Britain's current habitats – and, crucially, their interrelationships – must have existed for long enough, and over large enough areas, to enable the evolution of all the species we know today, as well as many that have been lost to extinction.

The analysis of ancient pollen preserved in waterlogged soils helps build the picture, providing evidence of an open but woody landscape in the period between the last Ice Age and when people began to farm; hazel, for instance, is abundant across parts of Britain in the pollen record but we know that it only flowers in full sun rather than in the shade of a closed canopy. Pollen samples also show that the composition of the wildwood was different to the treescape we're accustomed to, with small-leaved lime dominant across large parts of the south and east while a more familiar mix of oak and hazel was more common in the north and west.

Vera's thesis is obviously also strongly supported by all the species, like willow warblers, nightingales and purple emperors, that only really thrive where there's a complex structure of open space and different stages of woody growth: they must have evolved in a landscape with those features. There are lots of other species that need this diversity, as we've seen, not to mention all the species that clearly evolved in open areas, like the wildflowers of species-rich grassland. The saplings of many of our common trees, including oak, won't grow beneath the shade of a woodland canopy.

It goes beyond vegetation structure, too. Conservationists have long known about the reliance of many solitary bees and wasps, for instance, on bare ground in which to burrow tunnels; we're used to seeing their holes on paths worn across grassy fields by human feet or in the banks created by occasional minor landslips. But the recent reintroduction of big herbivores in rewilding projects across Europe has demonstrated the crucial ecological role of heavy hooves in creating this ecological niche at a meaningful scale, and paleo-ecologist Elena Pearce explained in a workshop online that these tunnelling invertebrates probably evolved in the bare earth at the side of 'bull pits', where male aurochs – and other ancient herbivores – wallowed and scraped to attract their mates.

Ancient trees and their occupants also play an important role in helping to understand the wildwood. Most British tree species will thrive in the open, developing spherical crowns as the branches follow the light out as well as up. These open-grown trees get much older than woodland trees, which typically grow close together and draw one another up before collapsing relatively young due to the impact of the wind on whole stands or their leggy vulnerability to decay. I was surprised at first to realise that we don't often find ancient trees in ancient woods; they represent distinct aspects of ancient landscapes.

Trees that co-exist for hundreds of years with wood-consuming fungi become host to those many saproxylic invertebrates that rely on various stages of decay, like the great oaks at Ashtead. What's more, the adult forms of many beetles and flies also feed on the flowers of species like hawthorn. Those invertebrates must have evolved in similar conditions – in old, decaying trees, and with scrub species flowering in the sun nearby – and entomologists have found plentiful preserved remains in deposits from the wildwood. Open-grown, ancient trees have therefore been around a very long time even in evolutionary terms, and they only grow to such an age in places like this.

Ancient wood pasture landscapes like Ashtead Common, then, have come to mean even more than their crucial relationship with human history would imply. Like coppiced woods, their long use by people unintentionally replicated the more natural, pre-agricultural landscape in which all species evolved, and probably even more closely. Their full ecological richness has similarly been lost thanks to their neglect over the last two centuries, as well as wider human impacts on nature.

As we respond to the crisis in biodiversity, the restoration of historic remnants like Ashtead should be at the vanguard of wider efforts to create more vibrant, ever-changing landscapes of open habitats, scrub and trees. Although nature conservation isn't about restoring the past, we can use our understanding of ecological history to ensure we provide the best possible conditions now for our threatened wildlife to recover and thrive. There's no one right way to do this. In some places it will be necessary to apply a human hand as we recognise the cultural significance of our interrelationship with trees since the Bronze Age, like in many of our ancient woods and on wooded commons. Elsewhere there's scope to hugely increase the area where we allow – and support – natural processes to drive change in rewilding projects, and perhaps to embrace more novel measures to restore the influence of long-lost species of megafauna on the surviving wildlife that evolved alongside them. Bison have been reintroduced to one British wood, as we shall see in a later chapter, but what impact might a herd of elephants have in driving complexity and restoring long-lost ecological niches? Whatever methods we use, it's essential that we restore the health of – and interactions between – the widest possible range of habitats, enabling them to achieve even more than the sum of their parts in a flourishing, natural kaleidoscope.

I moved on from Ashtead and South London in 2013, desperately sad to leave that committed group of rangers and contractors with whom I'd formed some of the tightest working bonds I've known. You never really leave a wood behind, though, and I visit Ashtead as often as I can. Shaun Waddell, the senior ranger I worked with then and who's still with the Corporation now, reminded me when we caught up recently that when the project began 'the old pollards were being suffocated by holly and young oaks and birch growing right up through them. Now, fifteen years on, each of the old oaks stands proud in its own space.' Since the start of the project in 2009, 900 of the 1,000 or so living oak pollards have been pruned, and every tree that needed to be freed from competition has had the work done, meaning the common is now more open than it has been since at least the mid-nineteenth century.

Just as importantly, by the summer of 2024 Shaun and his colleagues had managed to return cattle to about half of the common using electric fencing, maintaining the open structure kick-started by Jim and his team and restoring the wood pasture landscape to something resembling its historic variety and dynamism. The rangers cut sections of the rest on rotation with their regular group of volunteers, almost like giant coppice coupes, acting in the role of the missing herbivores while they negotiate the challenging logistics – and politics – of extending the fencing.

I'm looking forward to seeing the cows expand their territory and get up around the densest concentrations of oaks. Ideally they'll eventually roam the whole common, unconsciously dictating the progress of scrub, maintaining glades, and driving the constant process of change that will see wildlife thrive under a more natural system. Perhaps, with luck, some pigs or even boar will keep them company, creating with their rootling the conditions for new groves of sallow or other species of bare ground. When I used to check the cattle on those summer evenings they added the final touch to this landscape, shining dark red in the summer sun; they slipped easily between hawthorn and hazel, and tangles of bramble alive with insects, through the willows and under oaks, where they, too, seem more powerfully at home than they would in any field.

Most excitingly, Shaun told me that nightingales have been heard on the common for the first time in living memory, a sure sign that the rangers' work to restore the landscape is on the right track. Other birds of the scrub like yellowhammers, wheatears and even cuckoos – in catastrophic decline more widely – are breeding there in increasing numbers. Shaun feels confident that the numbers of purple emperors, too, are increasing with more open space and more willow scrub.

While I know in my rational mind that this restored landscape will be better for nature, there's also simply something in my heart that likes it, whether I'm gazing out from the top of an ancient bolling or walking among the willow scrub of the lower common. Humans have been called a species of the woodland edge, but what this recognises is that we, too, are a wood pasture species. The varied, woody landscape of a restored Ashtead Common echoes both the landscape in which we first evolved, in Africa's savannah, and the wider landscape in which we came to survive as we moved north and began our slow colonisation of the European wildwood. Forest scientist and writer David George Haskell explains that 'An affinity for savannah-like landscapes is one of the neurological quirks that we humans have carried with us as we spread across the world.'[6] Wood pasture has been home for people in Britain since our earliest history as hunter-gatherers, through the 'time out of mind' in which customs of communal use developed, and into the long period of formal common rights – and it only stopped being home when the Industrial Revolution and Enclosure broke our deep connection with the land. It's no wonder that it felt good to be back among it, working hard with other woodlanders towards a better future.

CHAPTER SIX

Woodbanks and Walking Trees

Woodland archaeology and immortal plants

EVERY BRITISH LANDSCAPE is a palimpsest, with the marks of generations of human use layered over one another in tantalising hummocks, shallow ditches and the patterns of its vegetation, and almost every ancient woodland is full of archaeology. In woods I've known well there have been Iron Age hillforts and mysterious enclosures like at Ashtead Common, and barrows on hilltops that once connected entire landscapes of Bronze Age memorial and ceremony. I've been shown hidden lines of giant stones that formed the foundations of medieval farmhouses, and many woods in the uplands host dramatic holes and depressions as a result of mining and quarrying. Perhaps even more resonant are the well-defined emplacements and rubble tracks left over from the heavy military use of British woods during the world wars of the twentieth century, particularly in the south-east in the approach to D-Day.

But, as woodland expert Ian Rotherham puts it, the

archaeology *of* ancient woodland is something different.[1] Testament to the way the wood itself was used by people, it provides vital and fascinating clues to a wood's history as a human artefact in its own right rather than as a load of trees that happen to grow on the ruins of some other monument. The archaeology of ancient woodland is unique in relying so heavily on ecology: beyond the woodbanks, charcoal hearths, trackways and ditches that hint at how busy with human life working woods must have been, our understanding of how people historically used woods relies on wildflower indicators as well as the tree species we find and their form. Gnarled boundary pollards and the spreading clusters of overstood coppice stools speak a language that's immediately relatable to a fellow woodcutter, revealing their last human interaction, its rough date and something of the intention of whoever did it; and I can easily imagine *how* they did it, and what it felt like using tools that are, for the most part, similar to those we use today. The stories told by trees like this go beyond that last cut, too, providing clues about a whole series of decisions made by our ancestors as they tended these woods for the things they needed, retaining or promoting the species we find today to the exclusion of others.

Small-leaved lime is perhaps the tree with the most intriguing story to tell, and it's a species that accompanied me on my journey as a ranger when I left Spring Park for the Lake District and Dodgson Wood. Delving into the history of both woods, and coming to know their rare, wild lime trees, revealed more to me about the industrial use of ancient woodland by people as well as offering a tantalising vision of an even more distant past. As living remnants of the wildwood, parts of these two woods are suffused with a sense of time almost beyond comprehension.

I wound up working in Dodgson Wood in the depths of a drab, mild winter, after having grown homesick for the hills of the north and tired of the aggressive pace of life even in the suburbs of London. I took a job with the National Trust in the Lake

District almost on a whim after the end of a long relationship underlined the need to move on, sad as it was to leave Spring Park and Ashtead Common. My new ranger role mainly took me into the valleys and towards the farmers I worked with but Dodgson Wood, a southern outlier of my patch on the shore of Coniston Water, had a grip on my imagination even before I moved: a pair of wiser woodlanders had visited Spring Park a few years previously and spoken in reverential tones of its charismatic population of lime.

Hugh Milner is a retired Forestry Commission woodland officer, a warm and gentle man who takes infectious, unselfconscious delight in trees. He was with Jonathan West, who was more reserved; Jonathan reminded me of Ben Law, the woodman who became famous after his house built from coppiced sweet chestnut featured on *Grand Designs*, and seemed to possess a similar combination of aesthetic appreciation and deep practicality. He lived in a wooden house that he'd built down in Hampshire, and worked the woods. They came to look at Spring Park's small-leaved limes, which I'd identified and understood to be unusual but whose true significance, at that point, I hadn't really grasped. Hugh and Jonathan were full of stories of lime – history and folklore, practical uses, mysteries and theories – and having seen the species through their eyes I began to fall under its spell.

The first lime we reached, strolling out of the tin-roofed office on a sunny June day, was a coppice stool with four chest-wide stems clustered together to form a huge crown, the result of the stool having been neglected for decades. While hazel, as an understorey tree, stays relatively low and bushy if it's not cut and just develops into ever-gnarlier tangles, the shoots of coppice-growth of canopy trees will naturally thin themselves as they grow, the stronger poles outcompeting the rest as they develop into a clump of mature trees emerging from a shared base. The trees that have grown large from abandoned stools are typically referred to as 'overstood' coppice, and this corner of Spring Park was one of two areas of the wood that weren't originally restored. The big, multi-stemmed limes were dotted among similar-sized stools of sweet chestnut and together they formed a dense and dark forest of tall, straight poles supporting a mature woodland canopy around the boundary between Spring Park and the neighbouring Threehalfpenny Wood.

Hugh and Jonathan's focus on the limes drew my attention more fully to the differences between the species and I realised that this first lime, standing as it did on the edge of the wood, emerged above the top of the surrounding chestnut by a good couple of metres. Its crown, even as a composite of the four stems, was narrow and elegant, an elongated egg shape that tapered to a rounded point and which seemed to flow in drifts and bulges like pillows of powdery snow: long, pendulous sprays of foliage dangled in shifting curtains. Its canopy on the field side was made pale by a cloak of yellow bracts, leafy wings that carry the seed on the wind and which stand up above the surrounding leaves in June in readiness for the flowers to emerge.

The leaves themselves are a delicate, slightly lopsided heart shape with a classy serrated edge that runs to a finely tapered point, and, while I'd obviously noticed them previously on casual inspection, I hadn't realised just *how* delicate they are. Most of the leaves up in the crown were only 5 or 6 centimetres across, and as we stood beneath the canopy looking up they became a sparkling filigree, light glinting through the shifting spaces between like sun reflecting on the ocean. The leaves stay lush and soft all summer – Hugh told me, in fact, that they can be eaten in salads – and even where the canopy was at its most dense the space beneath glowed a vibrant, comforting green as the sun passed through the thin fronds.

Hugh and Jonathan each plucked a leaf and turned it over, Jonathan pulling a hand lens from his top pocket to examine the hairs on the underside as he distractedly told me what they were up to: checking that the trees are definitely the small-leaved variety of lime. It was unnecessary here, he said, to look for the telltale tufts of microscopic red-orange hairs, because this population of lime is well known, but checking the leaves is a compulsive habit among lime enthusiasts, as I would come to learn.

There are two species of lime native to Britain: small-leaved lime and large-leaved lime. They're among our rarest native trees, generally restricted to tiny populations of a few specimens in out-of-the-way spots like this. (There are a few woods of pure lime in places like the Lincolnshire Wolds and the Malverns: they tell another aspect of the species' story.) There's also a nursery hybrid of the two, known as common lime and certainly the lime we find most often outside ancient woods. Common lime is, in comparison to small-leaved lime, a 'vulgar object', as Peter Marren wrote, its leaves less delicate in both form and texture, its canopy too dense; it is found only where people have planted it, most often in the ornamental parks of stately homes but also along Victorian streets.[2] As the names of the two native species suggest, the size of the leaves gives us a good clue to identification, but there's some overlap, so inspecting the length and colour of tiny hairs on the underside gives a more definitive idea.

We moved on past a handful more overstood coppice stools and, arriving at an informal junction of four woodland paths where Spring Park joins its neighbour, rounded the corner into a faint but unmistakeable gully, the round pebbles of the underlying gravel beds exposed in its bottom as it rose up the slope. The visitors stopped dead: 'Wow!' Two gnarled lime pollards stood before us, ancient trees that would have historically been cut just above head height every few years. Pollards are more often found in historic open landscapes like the wood pasture of Ashtead Common but Hugh explained that here, between two incontrovertibly ancient coppicewoods, the pollards served as markers; the woodcutters would have left them high when they cut the adjacent coppice to emphasise the boundary on which they sat. Sometimes you find them between coupes within a wood, but more often they sit between two woods that were owned by different people, like the pollards Jamie cut on the woodbank at the top of Forge Wood.

The bollings of the two pollards at Spring Park are typical in being intensely characterful: there are swellings and bulges among the rough sinews of mature bark and broken stubs of branches, and one has a gaping hole from the loss of a limb that leads to a dark pocket that disappears down the centre, full of stringy wood and scraps of cobweb. The regrowth from the top of each – two poles on the lower tree, three on the upper – is straight and its bark smooth, in marked contrast to the bollings, youthful trees perched on top of the decrepit trunks. We compared the size of the pollard poles to the surrounding coppice and it seemed evident that they were of the same generation and had, presumably, been last cut at the same point. It's impossible, Jonathan said, to judge how old the bollings are: they're relatively thin for old pollards, less than a metre across, but full of signs of age: 'At least a hundred and fifty – could be much more. I wonder if they marked this junction as well as the boundary. You know this is a *really* old boundary, right?'

I didn't. Hugh explained that the ditch that the pollards grew next to wasn't natural, but is almost certainly a centuries-old human creation; Oliver Rackham realised that many ancient woods are surrounded by the remnants of old banks – known as woodbanks – and their associated ditches, and that they're some of the most significant features in helping confirm a wood's age. I knew that the hollow here was the dividing line between Spring Park and Threehalfpenny Wood, and so also the boundary between modern London boroughs: Spring Park is in Bromley, while the neighbouring wood is in Croydon, and owned and managed by the borough council. I'd wondered whether something interesting was going on, but modern-day footfall can also wear big furrows along unsurfaced paths, especially on gravelly soils like this; Hugh showed me the signs that confirmed the antiquity of the earthwork as we wandered further up the hill.

'The ditch is wider than the path, look. And it's shallow, which usually means it's really old; it's been softened by age. But it gets really exciting here. We can be sure this is a ditch because there are banks either side.' He showed me the shallow humps running parallel to the path, lost a couple of metres back among the vegetation. Now worn low by time, the banks around many ancient woods were much taller when they were built and probably topped with a wooden fence; a ditch outside the bank would have further helped keep livestock out of the wood. Medieval deer parks, on the other hand, were enclosed by a ditch on the *inside* of a bank, to keep the deer in, and which was also sometimes supplemented with a stone wall. At Whiddon Deer Park, on Dartmoor, one of my favourite wood pasture sites with an inescapable sense of deep time, some of the boulders in the wall are the size of small cars and few of them are smaller than suitcases; the thought of them all being shifted and placed by horse and human alone is staggering. Even digging and building an earthen woodbank like at Spring Park would have been no small undertaking with hand tools, and the earthworks are testament to the high value of the woods they define and the historic importance of their protection.

In some woods, it's possible to quickly read the history of their extent by their woodbanks: where an old wood meets a field and there's a ditch between the field and a woodbank, we can be fairly sure this is its historic limit. Over the following months I'd realise that there's another hidden bank and ditch at Spring Park, running perpendicular to this one from the two pollards and marking the wood's southern edge. That bank matches the edge of the wood as it's shown on the earliest maps I've seen, and confirms it as the historic southern limit of Spring Park.

Other places can be much more challenging to read, where different eras of woodland management have come to overlay one another: we might struggle to understand if the ditch is inside or outside the bank or whether an adjoining wood is ancient or more recent, and it can be difficult to discern whether we're looking at the remnants of a deer park, a coppicewood or both, over time – or something else entirely. Like much archaeology, it comes down to interpretation and an element of guesswork, but we also have the ancient woodland indicator species to help. A carpet of wood anemone and celandine restricted to one side of a bank even within a wood, for instance, tells a tale that the flowery side is probably ancient while the trees on the other side have probably grown on land that was open more recently. Perhaps even more powerfully, in many old woods we can also read the species and form of the trees, some of which illustrate still longer timescales.

'The banks will be easier to see in winter,' Jonathan said, helping me discern their course through the foliage as we walked up the ditch, 'but we know they're really old because they're *covered* with lime.'

It felt like my perspective was being pulled into even sharper focus as I looked back: sure enough, there was a rough but unmistakeable line of lime trees on both banks. There were other limes off to either side back in the woods, and they were surrounded by hazel and holly, but the informal rows formed a definite feature that was suddenly obvious: there were overstood coppice stools, single trees and a couple of groups of young trees, culminating with the pollards at the junction at the bottom, an unmistakeable band of lime running along the boundary between the two woods.

'But what does the lime tell us about the age of the ditch?' I asked. Apart from the pollards, these weren't ancient trees; surely they could have grown here relatively recently having blown in from another wood. Hugh grinned and even Jonathan seemed excited as he told me: 'The lime won't grow from seed, so they can't colonise new places – they've been stuck round here for thousands of years.'

The pair explained that, since at least the Bronze Age, our climate's been too cool for lime seed to develop, as it needs a few weeks of continual warmth in midsummer: things are changing now, they noted. This protracted inability of small-leaved lime to set seed is the key to its significance and why many woodlanders find it so fascinating. The species survives by having an evolutionary trick up its sleeve: it is almost irrepressible in its adaptation for 'vegetative' reproduction, with individual trees springing so successfully back to life whenever they fail that the species is sometimes considered immortal. Oliver Rackham has called remnant limes 'a living link to the wildwood' and the presence of wild specimens is, for this reason, one of the strongest indicators that a woodland is ancient.[3]

Hugh pointed out a couple of examples of the species' remarkable reproductive strategies. We rootled around among what I'd perceived as a clump of saplings, and they turned out to all be growing from a low, broken stump like a natural coppice stool, parts of which had disintegrated completely back into the soil and leaf litter, and the rest of which was crumbly and soft, barely recognisable. A line of three youngish trees – each stem around the size of my thigh – caught his eye and we knelt to pull papery leaves and dry humus away from an indistinct hump running between them. Again, we found soft, moist white wood run through with the black 'bootlaces' of honey fungus, the remnants of a fallen stem.

'As long as some of the roots are still attached,' Hugh explained, 'they stay alive when they fall. Dormant buds in the bark on the underside develop into new roots and the branches on top go for the light and become new trees. Sometimes the original stem keeps going and the whole thing develops together, but sometimes it decays away, like this, and just leaves the row of new trees.'

I began, dizzyingly, to understand the implications of what I was learning: 'So these trees have been here since lime last set seed, just reinventing themselves like this?'

Hugh and Jonathan both smiled and nodded, and told me that the thickets of lime were probably the oldest trees in the wood, each one thousands of years old but in a state of constant transformation as they regenerate with each failure and creep along and around the old boundary – having wended their way back on to it from the surrounding woodland after it was excavated, with inconceivable effort, hundreds of years ago. The limes have persisted in this little corner of London since before people began to cut trees, witness not only to six millennia of human industry but also, probably, to the wilder time before that.

After Hugh and Jonathan's visit, I was a lime nerd too, and before I had even thought about working in the Lake District I knew about the trees in Dodgson Wood. They're famous, if that's the right word, among this select fan club thanks to their detailed study by the late Professor Donald Pigott, the global authority on the whole genus of *Tilia* – which includes the two native British limes, related trees across Europe and Asia, and species in North America, where they're known as basswoods. A couple of years after meeting Hugh and Jonathan I got talking to a friendly older couple at a woodland conference and, after they asked me where I worked, was astonished that they knew the Spring Park limes: 'And how are the old pollards at the junction faring?'

On finally introducing ourselves, I realised that this was Professor Pigott himself, with his wife Sheila ('He knows who you are, Donald,' Sheila cut in kindly when he began to explain his background). In time, Donald would show me his personal record of pretty much every remnant native lime in Britain. Half a dozen big white ring-binders held prints of photographs, many featuring Sheila or students for scale, along with handwritten notes of their location, leaf size, form and other details relevant to his quest to fully understand the species. Even in his early eighties, when I met him, he could talk about all of them in affectionate detail without recourse to the binders, as he had about the Spring Park trees when we first met.

At the conference I'd asked Donald about the way that lime had been pushed to the edges of ancient woods, like at Spring Park. He mentioned Dodgson Wood as his type-specimen of a population of small-leaved lime hanging on to survival, and evocatively described them lining the steep ghylls above Coniston Water. He pointed me towards his monograph on *Tilia*, a heavy academic tome with a global scope but featuring a hand-drawn plan of the lake with the streams running out from it like spiders' legs, and dots along them marking the limes that gave it the air of a treasure map. He told me: 'They're about as close as you'll get in this country to wild trees.'

With an increasingly crumpled photocopy of Donald's map in hand, I was able to kid myself that my first foray into Dodgson once I moved north had an element of exploration. I looked up the lake as I parked; the grey sky was almost oppressively open above the bald fells, the water pushed dull and matte by a stiff breeze. Tufts of mist clung to the big wood rising to my right like sheep's wool on barbed wire and I climbed a rotting field gate into the welcome intimacy of the trees. This place is special enough even without the lime, a classic Lake District oakwood that would now be classified as the 'temperate rainforest' that's captured the public imagination thanks to the work of Guy Shrubsole and others. Its proximity to the warming influence of the Atlantic and the Gulf Stream means it's relatively mild all year round, while it also rains – a lot. Combined with the clean prevailing winds, free from atmospheric pollution, the climate is perfect for lichens, mosses and liverworts, which, true to form, cloak every surface in Dodgson with lush gardens of life. There are deep beds of asterisk-shaped moss on every boulder and lichens on every trunk and branch, from minute crusts to big, green leafy specimens and everything in between.

I've never been able to get the lichens to stick in my memory – beyond the distinctive lungworts and 'stinky' *Stictas* that are the iconic species of the rainforest – but once, in a spare moment in the top of a rainforest oak, I counted fifteen different species in the last half-metre of one branch, in an extraordinary range of shapes and colours. The effect in winter is that, where lowland woods like Spring Park turn the brown of exposed trunk and leaf litter, the rainforests stay green, a pleasantly encompassing effect full of soothing texture.

Later in the year, in the most characterful bits of rainforest on the rockiest slopes, it becomes almost overwhelming in the best possible way, and I'm not sure there's a woodland experience in Britain quite like clambering around mossy lumps as the spring sun rises through twisted oaks so exuberant in form and so blanketed with other life that it can become disorienting; near and far seem to merge in the big tangle of green, and even up and down make less sense. They're always at their best just after rain when the lichens are engorged and their colours alive, the foliage and ferns dripping and shining in light that seems to be amplified rather than shaded by the fresh canopy. It feels like you can breathe more deeply, so powerful is the sense of photosynthesis from everything around you. And even when it's sheeting down, it feels like the natural order of things. 'Here in the rainforest,' Robin Wall Kimmerer writes in *Braiding Sweetgrass*, 'I don't just want to be a bystander to rain, passive and protected; I want to be part of the downpour.'[4]

On that grey day, though, I followed the first ghyll I came to up the steep slope among oaks that rise, in this part of the wood, straight and tall above a light understorey of hazel and holly. Occasionally I'd scramble up mossy scree and climb little outcrops of rock, and I clambered over a fallen oak and under a windthrown wych elm to stay close to the stream that tumbled to my right, its little gorge lush with hard fern and polypody. Then, putting paid to any idea I might have had that this would be a challenge, a chaos of fallen grey-black stems drew my eye among the surrounding oaks, splayed trunks sprouting thickets of whippy shoots as they arced across the beck. More big stems rose from the point where the tangle originated, on a nub of bare rock above a little cascade, and I traced them up to an almost geometric spray of fine, zigzag branchlets that, next to the chunkier twigs of oak, seemed ordered and refined. The biggest and most dramatic section of this rambling lime perfectly framed a platform hewn from the hillside, clearly the work of people but now camouflaged by the moss and ferns that had consumed it; the charcoal hearth was set almost like the stage in a theatre beneath its arch of spreading crown.

As I explored further over the coming months and years I realised that, by chance, that first tree is one of the most spectacular in the whole place. On finding it, I scrambled around to trace the connections between the bases of the various trunks, standing and fallen, and climbed among the chunky, sinuous roots flowing over and through the angular mudstone crag. It was an extraordinary thing in both its size and shape, and it's hard to believe that a worksite as serious as the hearth wasn't placed beneath the lime by design; it would have been a decent enough tree a century ago, with each trunk now a metre or so in diameter at the base. Its setting by the waterfall and the dark, brown pool below was almost comically picturesque.

One of the big stems had flopped over halfway up, its fibres unwinding like a giant hawser, and its top lay doubled over upstream. The thick, uncoiled strands looked like they should be soft and pliable to the touch, with bright wood exposed beneath the torn bark, but instead, of course, they were tough and bristled with splinters. I hunted around, like all lime nerds must, for evidence of the tree's regeneration: for branches rooted where the crown of a prostrate stem lay firmly on the ground, or simply where low boughs had dipped to earth and helplessly erupted new roots to anchor themselves and begin anew; but without success.

As I continued up the ghyll there were more limes, perhaps a dozen trees on the banks that rose to the surrounding oakwood or clinging to exposed stone that hung out over the water. The natural history of each was written in its form: short rows of three or four mature trees must surely have begun life as the side branches on the same fallen stem; others looked like overstood coppice, a cluster of poles from the same stool. (Nobody's really sure, I would find, whether some of these survivors were coppiced or whether they're trees that failed naturally and sprang back from their broken stumps; Professor Pigott believed that most had never been cut.) Some of them had been there so long that the banks had eroded and their roots created a scaffold that suspended them out over the stream.

Over the next couple of years I returned to Dodgson as much as I could. There was no felling planned during my time in the Lakes, but other work took me there: checking and fixing the boundary fence, ecological surveys, or meeting Natural England staff about its designation as a Site of Special Scientific Interest. I visited more often on my own time, when I left the yard at the end of the day or cycled over the hill from Ambleside on a Saturday morning to make my way up and down each ghyll in turn, the shaky, semi-parallel lines on Donald's map coming to life as I beat new paths alongside them. I was lonely and bruised by the break-up that had prompted my move and, outside work, I needed something to draw me to the woods, where I'd never regret going if I could just get started.

There was something particularly restorative in sitting beneath the limes, my ears filled always with the white-noise rush of a beck, and contemplating their unimaginable persistence. I've always been prone to feeling low but, when I made the leap into conservation in my early twenties and started to learn about the countryside and the natural world around me, I slowly developed a sense of peace and belonging I'd never had before, except perhaps in the solace I found in the woods as a child. It took years, but I began to feel at home with myself and my place in the world, taking deep comfort in knowing the trees, birds and flowers, and in starting to understand something of the history that has brought these places to be how they are. Physical work helped, too, and collectively it felt like this was the stuff I'd been missing. That's not to say I'd been riding a wave of elation since then, but my new relationship with nature and its conservation certainly played a part in my coming to a kind of accommodation with the melancholy that forms a part of my psychological make-up.

I realised, a decade on, that I had perhaps begun to take this quiet, life-changing discovery for granted and become accustomed to a new baseline. Uprooted as I felt by the end of the relationship in London, it took time to remember what it had felt like to have the natural world revealed by Darren, the ranger in Cheshire who took me under his wing, and the other rangers who'd mentored me early on. Heading to Dodgson didn't always do it, and at first I'd feel even worse if I went to the woods and didn't go home feeling noticeably lighter; I had to relearn that it's a long game. Each trip out was good for me in some way, though – I usually saw something interesting, or learned something about the woods, or about myself – and their cumulative effect, along with some high times with a little crew of friends who got into the habit of visiting, was to bring me back to life. Like the walking limes, I came to terms with the fact that I wasn't quite the same shape I'd been before, and never would be again – but that I'd grow into a new version that would persist, for a while.

I'm not blind to the fact that I became something of a walking cliché, at risk of becoming what Kathleen Jamie memorably described – with justified impatience – as the 'Lone Enraptured Male', off on my self-absorbed spiritual quest.[5] Being generous to myself, though, I perhaps came to understand Wendell Berry feeling, on considering the age of the woods on his farm, 'like a flea in the pelt of a great living thing, the discrepancy between its life and one's own so great that it cannot be imagined'. Sometimes, when the soft, heart-shaped leaves began to emerge over a scattering of wood anemones and they all rippled and shifted in the spring breeze, I would lie in the leaf litter and feel like I might stay there for ever, comfortably consumed by the moss and soil into the woodland itself. Rather than being nihilistic, it was as near as I've ever come to genuine, thoughtless peace; what Berry called 'the spell of primitive awe, wordless and humble'.[6]

Donald's remark that the Dodgson limes might be the closest we'll find in Britain to wild trees stayed with me. By his reckoning, the trees here have been left even more to their own devices than those at Spring Park, some of them having walked the ghylls undisturbed throughout the whole course of human history. Studies of pollen records – including important samples from waterlogged flushes at Dodgson Wood – have shown that small-leaved lime was one of the most populous tree species before humans began to modify the landscape, dominating the canopy of the wildwood across large parts of the southern half of Britain. Its gradual decline probably began as soon as people started to heavily influence their surroundings. Unlike most other British trees that – given the chance – readily recolonise places from which they've been lost, the removal of lime was irrevocable due to its inability to set seed.

The wood and timber of lime wasn't as useful to our ancestors as species like oak and ash as it's relatively soft and light and, although it made good shields and could be easily carved into bowls and utensils, it burns quickly and without much heat. The tree was, however, vital to people from the Stone Age until the Middle Ages as a source of cordage; its fibrous bast – the inner bark – would be 'laid up' into string, then braided into rope.

Lime's fibrous nature, unique among British trees, and its subsequent significance for people until just a few hundred years ago was impossible to ignore as I came to know the species; in many indigenous languages across the range of *Tilia*, limes are called variations on 'string tree'. When he came to Spring Park, Jonathan found scraps of dry bast peeling from a dead branch and separated out a tan-coloured strip a couple of millimetres wide and as long as his forearm. He held either end between thumb and forefinger and twisted until it started to contort, then pinched the middle and folded it in two. The threads twisted around one another and stayed in place to form a short but unmistakeable piece of twine. He explained that not only are the fibres of the bast long and tough but they also hold this double twist, which is, he said, the defining property of most materials used to make thread. I still can't help but fiddle a little bit of yarn whenever I find bast peeling from a dead lime twig – which I do often, as lots of streets near where I live are lined with common lime. After a stormy night, wind-felled twigs are crushed by the wheels of passing cars and the bast reveals itself and is easy to peel and twist, to the complete disinterest of my children.

More recently, I met Dave Watson, one of the longest established bushcraft trainers in Britain and an expert in replicating the ancient uses of wood. He told me how he strips the bast from in-cycle lime coppice while it's still standing, nicking through the bark at the base with a knife and then simply pulling it back as high as it will go in a series of strips, which he then severs once the pole is felled. There's a narrow window in June when the sap has risen and the bark comes away easily in long ribbons that produce the highest-quality string. The sheafs of bark are rolled up and 'retted' for a few weeks, submerged in water so they decompose slightly; the bast then falls away from the outer bark more easily, and softens for use. Golden coils come out of the water stinking unpleasantly of pond-mud with an unwelcome sweet undertone of cider or straw. (I've tried to make lime string myself, although without the skill that Dave would demonstrate; if nothing else, the particular smell has stayed with me.) Once dry, he strips the bast into thin fibres, like Jonathan's but 4 or 5 metres long, and ties two to something solid – in this case, his truck.

Dave showed me something approaching magic as, starting where they were tied off, he gave the fibres a quick twiddle each, one per hand, then brought them together; they wrapped round one another and stayed in place as he twiddled the next bit, his string, made with the relentless speed of an expert knitter, rapidly starting to look even more like something you could buy in a hardware shop. The same process can be repeated with two or three pieces of string to lay them into rope, or they can be braided. When the Roskilde Viking Ship Museum sent one of their preserved boats to the British Museum, I was delighted to see among its accompanying artefacts a scrap of lime-bast rigging from around the tenth century (alongside rope plaited of horsehair, which must have been more time-consuming, and from sealskin, which I assume smelled even worse).

Like modern coppice-worker Brian Williamson observing which colour hazel stems make the best hurdles, Dave has also rediscovered lost knowledge that would have been commonplace to our ancestors. He noticed how different parts even of the thin layer of bast behave differently when he's laying them up; the inner bast makes the neatest string, while the outer makes it lumpier and more uneven. Dave visited 'Ötzi' in the South Tyrol Museum of Archaeology, the Bronze Age man mummified beneath a glacier until he was exposed by the melting ice in 1991. In conversation with the curators about the lime-bast string found among Ötzi's possessions, Dave was able to disabuse them of their assumption that the rough twine was the best that Bronze Age people could do and argue that this was simply the cheap stuff; it was well within the high level of sophistication demonstrated by the rest of Ötzi's possessions to make good twine. Ötzi's bag and shoes are made of woven lime bast string, and the shoes also contain bear leather and deer hide, with straw and moss for insulation. They were so difficult to recreate using flint knives and bone needles – and bast supplied by Dave and woven by another expert, Jacqui Wood – that archaeologists assume that specialist cobblers would have existed in the Bronze Age to make them. Upon trialling the boots in the snowy mountains the team also found them to be warmer and to have better grip than modern shoes.

It's hard to appreciate how important such a strong and versatile source of fibre must have been since its discovery at the peak of lime's dominance in the British canopy during the late Stone Age. Archaeologists have found preserved lime-bast cord from this period, as well as fishing nets tied from it, and the availability of strong rope went on to be quietly integral to much higher-profile technological innovations like sailing and ploughing. Ötzi wasn't alone in wearing shoes of lime, which continued to be common in parts of Europe until well into the twentieth century, although these were more typically sandals rather than insulated mountain boots; the woven bast was also used to make other clothing including waterproof capes.[7] From the fifteenth century, however, travellers to Asia began to bring to western Europe increasing quantities of hemp and jute, both of which made stronger rope and triggered the final, dramatic decline of lime in British woods. The species probably fared worse than any other in the dramatic changes two or three hundred years ago when less economic trees were cleared to make way for commercially managed coppice.

Where Spring Park was modified from its more natural combination of species to hazel under oak standards, large parts of Dodgson Wood were converted almost exclusively to oak, along with the majority of woods in the uplands of the west of Britain. The trees were grown as coppice rather than standards and the wood used to make charcoal, while the bark went to tanneries. Lime, now all but useless to people, was pushed to the edges of both woods as it disappeared from the British landscape, unable to recover and repopulate the woods from its tiny remnant populations.

The conversion of Dodgson to this heavy industry of oak is writ unmistakeably across swathes of the wood away from the ghylls; the hillside is home to more than 150 imposing charcoal hearths, among other woodland archaeology. Dodgson Wood's very survival is a result of the dependence of local industry on a sustainable supply of oak. Some of the hearths are so highly engineered as to feature retaining walls at their lower edges, lost beneath the fern-dotted moss; the hearth beneath that first and most dramatic lime is one of the widest and flattest I've ever seen, at least half the size of a tennis court. A couple of them are accompanied by the stone foundations of permanent huts that may have been home to the woodlanders, although from Victorian illustrations it's believed that the charcoal burners – or wood colliers – built temporary wooden shelters. There are, though, C-shaped rings of stones a couple of metres wide that are thought to be bark-peelers' huts; the peelers would have re-roofed them with wood and turf each time they returned to a particular spot on their decades-long cycle of cutting, as they coppiced haggs of oak and stripped the bark for the tanning industry.

The worksites are connected by two dozen tracks that criss-cross the slope, mostly lost but glimpsed in fragments of mysterious stone revetment or bulging crags chipped flat to allow the passage of carts. Like many upland woods, there's no woodbank around the perimeter, which would have been all but impossible to dig and build from the thin soils over bulging bedrock and scree. Once you get your eye in, though, the flatter areas, blasted crags and crumbling retaining walls become irreversibly apparent. Although there are charcoal hearths and old tracks in many ancient woods, I don't know another with such dramatic and well-preserved remnants of woodland industry and such a tangible sense of our ancestors who worked and lived there.[8]

Unlike some other woods, we also know that Dodgson was still bustling with activity as recently as the 1930s. It rises above Peel Island, the inspiration for Wildcat Island in Arthur Ransome's Swallows and Amazons novels, which I went through phases of being obsessed with as a child. Ransome enthusiasts believe that it's where he met real charcoal burners who inspired Old Billy and Young Billy, blackened and feral but friendly to the Swallows despite their notorious adder in a tin. Their burn took place beneath a mound of earth rather than a steel kiln, sods and turves arranged in a cone over the pyrolising wood and needing even greater continual care to ensure no oxygen got in to fuel a fiercer fire; by Ransome's account, one of the wood colliers was in continual action on the mound, each jet of blue smoke smothered with a shovelful of soil prompting another to begin. Hazel hurdles, in yet another use, would have formed adjustable windbreaks to help control the process, while the location of the biggest hearths by the ghylls is no coincidence as gallons of water would have been needed to extinguish each burn. You can scrape away the leaf litter on Dodgson's hearths with your boot and find a layer of scorched soil and fine charcoal dust.

From the Middle Ages charcoal was hauled down to lakeshore bloomeries, where it was used to smelt iron; the monks of Furness Abbey who first controlled the industry found it most efficient to bring iron ore to the wood by horse and boat, rather than transporting the bulkier, fragile fuel elsewhere. Jamie Lund, the longstanding National Trust archaeologist in the Lakes, one-upped my pursuit of charcoal seams beneath woodland hearths by showing me lumps of medieval iron slag in the earth around the bloomery sites, themselves marked by hummocks that only become apparent with his expert interpretation.

The industrial use of the South Lakes woods gathered pace with innovations in smelting. The availability of ore and the large area of remnant woodland sustained one another, turning the area into a thriving hub of both iron and woodmanship even as coke began to fuel the process elsewhere – much as the same industry kept Jamie Simpson's wood alive in the Sussex Weald. The development of bloomsmithies, with big water-powered bellows, required even more charcoal, while the unmissable stamp of industry in Dodgson Wood – the reveted hearths and quarried tracks – probably dates from the eighteenth and nineteenth centuries when smelting was rationalised into a couple of big blast furnaces that drove demand for charcoal yet higher, including one at Nibthwaite at the foot of Coniston Water.

The construction of the blast furnaces coincided with an increased need for tanbark, and that process also became industrialised in tanneries in Low Furness, down by Morecambe Bay. Oak became increasingly attractive as it paid twice, which almost certainly contributed to the main period of modification here where other tree species were selected out of the woods; the production of oak became so lucrative at this point that more trees were planted on sheep pasture in some places, including around the edges of Dodgson Wood. Illustrating the comprehensive range of uses for coppicewood, the furnace at Nibthwaite was also home to a bobbin mill – probably using the lower proportion of other species that were harvested alongside the oak – while the twiggy tops were burned with bracken to make potash, which went into lye to wash the wool of the Lake District's Herdwick sheep before it was made into clothing.[9]

Even as the Furness iron industry declined in the nineteenth century with the development of more efficient coal-powered processes elsewhere, its infrastructure was repurposed for new uses of the abundant underwood available nearby. The bobbin mills persisted and the mill at Stott Park, just over the hill from Dodgson, is now an evocative museum in the care of English Heritage, having processed its last load of coppiced wood in 1971. It's fairly obvious as you travel around the wooded valleys of the South Lakes, but the ancient woodland data on Magic Map confirms it: Dodgson Wood forms part of a massive concentration of ancient woodland between Coniston Water, Windermere and Cartmel to the south, unusual in both a Lake District and a national context. Like its fellow heavily wooded landscapes in the Sussex Weald and the Forest of Dean, the survival of the South Lakes oakwoods and their culture of coppicing is intertwined with the history of iron and the reliance of heavy industry on a sustainable supply of wood.

Even this history, of lime and oak, is too simple; like most bigger woods, Dodgson has expanded and contracted continually with changes in economics, the trees allowed to colonise the surrounding fells when the price of wool was low or that of underwood high, then nibbled back when it made sense to the farmers to graze bigger flocks. Its current boundary – which can seem so definitive to us – is actually fairly arbitrary, a result of the negotiated location of a fence that was only erected in the 1980s.

In the years between the Second World War and that point, with coppicing abandoned, woods like this across the hills of the west of Britain were seen as nothing more than places for sheep to shelter and were left open to the surrounding pasture, another symptom of Rackham's Locust Years. Professor Pigott, early in his academic career, created and monitored a trial 'exclosure' in a similar wood in the Peak District, fencing out a plot that proved, by 1983, that it was sheep nibbling all the saplings and preventing the growth of a new generation of trees, eating the future of the woods as they doomed them to eventual disappearance.[10] The National Trust and other organisations built fences around many of their upland woods at this point, Donald's study having reinforced the increasing awareness of the ecological and cultural value of ancient woodland. This unglamorous intervention saved many of the remaining areas of ancient woodland, at least, from the period of subsidy-driven overgrazing across the wider landscapes of the Lakes, the Peak District and Britain's other hilly places.

Within the wood, too, there is a whole series of layers of history to read in its archaeology and the species and form of the trees, all of which illustrate a constant process of change in response to the developing needs of people. On the lowest slopes, near the road, much of the wood is dominated by oaks that are straight and regular; they might have been 'singled', where one coppice pole is left standing to convert a stool into a timber tree, but it's more likely that they were planted in the Victorian era, when the value of oak as timber began to overtake its value as coppice. A tall stone wall runs uphill through the wood near its southernmost end, a clear indication that one side was grazed; the trees beyond it were almost certainly planted on an open field. Further up we find big groves of overstood coppice while, in some places, the oak is joined by alder and ash, the mix of species the mark of an older regime and of areas that escaped modification, perhaps because they were less accessible.

There's a patch of birch the size of a football pitch, dense but airy and with a wholly different atmosphere to the oak, where a plantation of Antarctic beech and larch was removed within the last couple of decades; it was planted by the Trust before the significance of places like this was understood, a classic example of PAWS – a Plantation on an Ancient Woodland Site. The natural colonisation of the birch, nearly always the first tree to exploit a big clearing like this, was enabled by a tall surrounding fence to keep Dodgson's resident red deer off the saplings. At the top of the wood as it peters out on to the bare fell, with sheep gathered along the wire, the oaks are at their most charismatic, squat and wind-hewn as they perch on boulders in scrubby, open woodland that conforms most closely to the popular image of temperate rainforest, and which is all the more appealing for the giant views across the canopy to Coniston Water and Peel Island, far below.

Nicola Chester writes that 'a landscape doesn't forget its stories. It wears them like lines on an old face, markings on an old body.'[11] People sometimes talk about thin places in a spiritual sense, as portals between worlds, but these unmistakeable layers at Dodgson make it a thin place in time, as if history is compressed into a moment and I can move between eras with a step – or a primate scramble up a little slope or quick fight through bracken and bramble. There is an entirely benign presence emanating from the hearths and hut circles, as if I'm accompanied by the families who lived and worked here for generations, filling the wood with sound. I sometimes expect to hear the solid thwack of an axe, a song, or the shouted warning of a tree about to fall over the call of a wood warbler – a sound that would have been familiar and particular here for all those woodlanders, as it is for me – or to put my hand down and feel the black earth still warm on one of the big hearths. And simultaneously, in the ghylls, the limes bring the pre-human landscape to life, showing us – more than showing us, helping us *feel* – the wildwood, these very trees having perhaps sheltered aurochs or wolves during previous iterations of their endless transformation.

The received wisdom is that the limes clung on along the ghylls because it was simply too difficult for the woodlanders who modified the rest to remove them from the crags and slopes; they were pushed as far as possible into the margins to make way for more profitable trees. I'm not so sure. George Peterken told me that he thinks the way unwanted trees were cleared before mechanical diggers came along was by the use of grazing: the coppice was cut, then livestock purposely penned around it to nibble the lush new growth until the stools died. Anyone who has ever uprooted and removed even a small garden tree knows that this makes a lot more sense than the common assertion that redundant trees were 'grubbed' out, which would have taken enormous effort. I think that a determined woodcutter would have been able to fell most of these limes, and I'm sure that the nimble Herdwick sheep of the Lakes would have got to the regrowth. Perhaps the trees were allowed to stay for a reason.

At Spring Park, the case is even more compelling for there being something else at play to influence the survival of the limes on the boundary, as Hugh told me as we brushed the leaves from our trousers and stood up from inspecting the hidden connections between the thickets on the woodbank: 'There's more going on here; Donald discovered in his research that before the London boroughs existed this was the boundary between Kent and Surrey, and *that* dates back to the Anglo-Saxon kingdoms of Kent and Wessex. This place has been an important line on a map for at least a thousand years, and the bank could be just as old.'

This astonishing history, I know now, helped explain the width of the ditch with its banks either side, but it didn't necessarily explain the survival of the lime, which it would have been easy to cut and graze on such a gentle slope. Hugh and Jonathan told me that lime continues to feature in the folklore of northern European countries like Germany and Sweden; 'linden' symbolises peace and wellbeing, and old trees often form traditional meeting places. They hypothesised that perhaps the lime was integral to the ancient boundary, retained along its length as a marker of peaceful co-existence between the kingdoms, or an aspiration at least.

It's an evocative, romantic thought, neighbouring Anglo-Saxon tribes meeting in truce on the lime-lined bank. But most crucial to the survival of the lime were the decisions of the woodlanders who've cared for Spring Park since then, in the fifteenth century when the value of the bast began to wane and, more dramatically, within the last three hundred years or so when the rest of the wood was converted to its near-monocultures of hazel and sweet chestnut. In the decade or more since Hugh and Jonathan first got me interested in lime I've been fortunate to travel to lots of woods, particularly in recent years in my advisory role with the National Trust, and this is a theme that I see time and again, although I've never found a written reference to it. At Leigh Woods in Bristol, among areas of both historic wood pasture and coppice, relic populations of small-leaved lime are restricted to strips alongside ancient boundary banks, helping reveal the geography of the old coupes; at Blickling in Norfolk, giant native limes that originated as coppice mark barely perceptible woodbanks that were later incorporated into the design of the landscape park. Perhaps there was some specific local demand at these places for the tree's fibre, or perhaps beehives were kept among the thin stands of lime to take advantage of their abundant flowers. It's thought that lime was often retained for this reason in other parts of Europe, and in high summer you can hear the insect buzz of a sunlit lime before you see it.

Maybe, though, there was a less prosaic meaning attached to lime that we have, in Britain, lost, even in the past couple of hundred years. Something perhaps to do with its symbolism of peace and good health – as well as fertility and love – that persists elsewhere in Europe; or maybe simply its beauty and sensory appeal. Professor Pigott asked a forester in Switzerland why he'd left a group of large-leaved limes alone while felling the woodland around them; the lime is, he was told, 'ein Lieblingsbaum' – a beloved tree.[12]

CHAPTER SEVEN

The New Dark Age

The sad story of woodland bats, birds and butterflies

WE FELL INTO silence as the sun dropped behind the trees; it's all but impossible to speak in anything more than a whisper at dusk in a wood on a clear summer's day. The sky dimmed through deepening shades of blue above a rich band of pastel yellow, while the backdrop of hazel, a verdant tangle, condensed into a stark silhouette. A blackbird punctuated the stillness of Spring Park as it dashed between branches with a chuckle before settling. We waited.

For most of us, spotting bats in Britain takes patience or luck – sometimes both – and you need to be in the right place at the right time; the same goes for many of our woodland birds and butterflies, the main groups of species that ecologists monitor to understand the health of woodland ecosystems. But it wasn't always like this. Until the Industrial Revolution and the diminishing significance of our woods for human essentials, they thrived with species that relied on the dynamism and structural diversity provided by coppicing. The years since the Second World War were not only Rackham's Locust Years but also marked the onset of a new dark age for those woods that survived as they fell into neglect, causing catastrophic declines in much of their wildlife – albeit after a little-known twist in their tale during both of the twentieth century's global conflicts.

Understanding the sad story of these key groups over the last few decades is vital in helping us restore our woods to the healthy ecosystems they were throughout centuries of human use. There are, though, places where woodlanders have successfully restored nuanced active management, by both traditional techniques and more novel approaches; in some woods, their work is already paying off in recovering populations of rare species, while in others we're seeing an equally important resurgence in the abundance of more common wildlife.

We'd all been out bat-spotting plenty of times before and were ready to be patient, but the quiet of the enveloping shadow still stretched long enough that Jamie and I were lost in contemplation; we both shouted with surprised delight at the eventual chatter on our detectors: 'There's one!' Mike, a professional ecologist we'd called in to provide some real expertise, was calmer, smiley behind his gruff exterior and bearded long before it was trendy; he looked a bit like he'd stepped out of Tolkien's Mines of Moria.

'Yep. Pip – went past me towards you, Luke.'

Pipistrelles are easily the most common bat in the UK, two almost-identical species that comprise 80 per cent of the bat population. They're typically the bats you see if you're lucky enough to be able to watch them from your garden, circling and diving as they repeat a route along the eaves of the house; and in woods, too, they stick to the open spaces – this one was using the slight gap in the canopy created by the paths. I'd become used to hearing 'pips' on the detectors, little black boxes of electronic wizardry that convert bats' ultrasonic echolocation calls to a pitch where humans can hear them. Jamie and I had basic ones; the different species of bat have very different calls and to identify them properly with our simple kit we had to tune the detector to the different frequencies with a little plastic wheel – not a simple task as the bats buzz past – and compare the pops and clicks we heard with descriptions on a sheet of card. Mike's detector was much fancier and he could listen to every frequency at once, as well as digitally recording it all for analysis later.

We were stood around one of Spring Park's big oak 'standards'. It had been dead for some time, but we'd made the decision that it needed some work to ensure the safety of walkers on the two busy paths that met beneath; the spreading crown was weighting the tree towards them and, once dead, it was destined to collapse at some point. Its bark had come away and was hung loosely in dusty sheets from the trunk, leaving a slim void that bats might crawl into to roost, so I'd called Mike to help make the tree safe while ensuring we didn't disturb them if they were there, as well as preserving their potential sleeping quarters. He's an ex-ranger who could be relied upon for pragmatic advice, and he recommended that we just take off the branches over the paths, leaving the trunk with its 'roost features' intact but with the mass of the remaining limbs meaning it would eventually fall harmlessly back into the wood. To avoid bothering any bats we'd do the work in early September when they've raised their young but aren't yet hibernating, carrying out an 'emergence' survey on the tree at dusk and a return survey the following morning. We'd then work on the tree immediately if no bats were roosting in it.

The pipistrelles looped in figures of eight and manic circles in the space above us and we could just make them out against the darkening blue, watching as their sudden ducks and dives coincided with rapid-fire zips that punctuated the Geiger-counter clicks from the detector: the rate of the bats' calls increases as they home in on a midge or fly. Each pipistrelle eats around 3,000 every night, so they do it a lot. After that first one, they became fairly commonplace, a couple coming by every minute or more likely, Mike told us, the same few patrolling a route. We checked in each time, whispering instinctively, confirming their trajectories and – most importantly – agreeing that they weren't coming from the oak.

As it got darker we heard a deeper, more regular double-crunch on our detectors: 'Noctule – along the top path.' Mike calmly shone his torch up just below the branches of the surrounding oaks and we caught a moment's glimpse of a bigger bat heading purposefully along the widest path in much more direct flight than the pips. Noctules are the largest British bat and, although I knew their wingspan is not quite as wide as that of a starling, by the fleeting light of the torch it looked as big as a crow and the hairs on the back of my neck stood on end.

An hour or so after sunset things were quietening down. Mike told us that standard practice is to stop surveying then, after the bats' most active period, and that if any had been resting for the day in our oak they'd be out; the bats rush from their main roosts as soon as it's dark enough to feed safe from their own predators, then rest in temporary roosts for the night, heading out occasionally in pursuit of more insects. We were walking back to the yard – fortunately with our detectors still on – when we heard a third style of echolocation call, much quieter and somehow also softer in tone, less defined. Just as we heard it all three of us became aware of something between us at eye level, passing within a couple of feet with an almost imperceptible *thurr*. The hairs on my neck went up again and it was impossible not to flinch in the pitch dark. Even Mike was animated: 'Brown long-eared!'

They're not rare – at least, not that rare, by bat standards, he said, 'but they are cool. They call much more quietly and hear the echoes with those big ears; it's an evolutionary arms race with moths that learned to evade other bats by listening for their calls and dodging them.' While pips and noctules use the openings between the trees, long-eared bats hover around the space beneath the canopy, their echolocation and hummingbird-like manoeuvring so precise that they can pluck a stationary moth from a leaf.

It's a minor, daily spectacle in summer to catch those half-visions of the bats' acrobatics and to eavesdrop into their busy language, a glimpse into a nocturnal world that reminds me that the whole wood is bustling with life beyond the edges of our meagre human senses. It's sobering, though, to compare the scale of this modern experience, spine-tingling as it is, with what's been lost: although there were no formal monitoring schemes until the last couple of decades, it's widely recognised that bat numbers plummeted between the Victorian era and the end of the twentieth century, and the dusk air in a wood like Spring Park must have teemed with life far beyond anything we experience today.

Like most other British wildlife, our nineteen bat species – almost half of the native British mammals, which number no more than fifty – evolved in the wood pasture landscapes of the pre-agricultural wildwood, all of them reliant on trees and woodland for some aspect of their ecology. Many species, like the pipistrelles, use open areas between trees or the edges of woodland to forage for insects. Other bats, like the brown long-eared that whispered among us in the wood, feed in the canopy itself. Part of the reason that pipistrelles outnumber other bat species is that they've been able to adapt to the loss of their arboreal roost sites by moving into buildings; the rarer bats, like the Barbastelle and Bechstein's bats, haven't been able to find human-made alternatives to the roosts they now struggle to find in old trees, typically in woodpecker holes and cracks. They often need a few decrepit trees in the same wood, and also have more demanding requirements of their hunting grounds. As with the pipistrelles, complexity is key – that magical combination of trees of different ages, scrub and open space – but their needs are more specific around the density of the understorey and their use of other woodland features like streams and ditches.

The clearance and simplification of wood pasture landscapes over the last few centuries and the removal of trees from farms forced many bat species to begin a gradual retreat to woodland strongholds, where traditional coppice management continued to provide similar conditions, with its varied structure and the warmth and boom of plant-life in the fresh coppice coupes supporting an equivalent glut of insects. But at the same time as woodland management fell into decline, agriculture became even more intensive, with changes like the reduction of haymaking and an increase in the use of pesticides radically depleting the insect fauna on which bats relied, contributing further to their diminishing numbers. The polarisation of the British countryside into dark, shady woods and relatively tree-less (and chemically treated) fields also affected the places bats roost and meant that they lost their navigational waymarkers in field trees and hedgerows.

Monitoring bats, along with butterflies and birds, provides a powerful insight into the wider health of woodland ecosystems. The size and diversity of their populations tells us about the status of the web of plants and insects on which they rely, while they're also relatively easy to identify and count. Birds and butterflies have been studied in a consistent way for longer than bats have been, with monitoring dating back to the 1970s, and both groups have experienced equally depressing downward trends, and for similar reasons.

Woodland birds have been recorded to a consistent methodology for more than fifty years and these records illustrate shocking declines, with data released by the government demonstrating a 35 per cent decrease across the UK from 1970 to 2022 in the woodland bird 'index', a combined measure of the abundance of thirty-seven species that use the woods. Twenty-five specialist species that rely, in theory, on woodland – like willow warblers and spotted flycatchers – collectively fared even worse, declining as a group by around half.[1] (As we saw at Ashtead

Common, although some of these species have been clinging on in the 'refugee habitat' provided by woodland they might be better adapted to healthy wood pasture landscapes. But we still need to make our woods better for them, too, and the lessons we're learning from wood pasture can influence that improved management.)

The declines in the specialists illustrate the wide range of factors that can affect British wildlife. Long-distance migrants like nightingales and wood warblers, for instance, might well be struggling due to changes in their summer homes of British woods – but the picture is complicated by the fact that there are also threats on their routes to and from Britain, as well as in their wintering grounds in sub-Saharan Africa. Desperate declines in willow tit, meanwhile, are due to the continuing loss of its rare old-growth wet woodland habitat. But the declines in many of the species studied are attributed primarily to the end of traditional woodland management and the subsequent increase in shade in both former coppicewoods and infilled wood pasture, as well as the loss of trees from farmland: lesser spotted woodpecker and spotted flycatcher join willow tit in the dismal group whose numbers have fallen by more than 90 per cent since 1970, and they both require large areas of open wooded habitat.

The same applies for birds like the garden and willow warblers, which also feature on the list of declining woodland specialists and represent the species that require the constant, ephemeral cycle of young, dense, shrubby places in which to nest, as well as open areas in which to feed on a thronging insect population, a combination simply not provided by most of our dark, neglected woods. It is important to acknowledge that some woodland specialists, like blackcaps and nuthatches, are on the increase, with the populations of these two species having almost doubled since 1976; the reasons for these changes are complicated and not fully understood. The fact remains, though, that overall woodland species are in serious trouble – second only to the farmland specialist birds, which have declined by 60 per cent since 1970, the intensification of agriculture again being to blame.

My initial reaction to the plummeting graphs displaying this miserable information was to compare them with similar ones showing the earlier dwindling of coppicing: according to government records, the area of coppice under active management fell from around 230,000 hectares in 1905 – when the practice was already in decline – to less than 24,000 hectares by 1995.[2] (The same records show that the majority of the woodland still in coppice management even by 1965 was non-native sweet chestnut in south-east England; while this still brings lots of the wildlife benefits of coppicing native species, it's not quite as valuable for nature as the management of species like hazel, and tells a different human story.)

Landscape historian Charles Watkins reminded me however that, while coppicing rapidly tailed off as major industry earlier, lots of ancient woodland was also used in some way during one or both world wars. These additional interventions prolonged – and in some cases temporarily increased – the provision of complex and open wooded habitat, and add an intriguing coda to the story of the use of our ancient woods.

By the outbreak of the First World War, Britain had come to depend on imports for around 95 per cent of its timber needs, while coppicing had all but died out due to the Industrial Revolution and drastic changes in the rural economy, as well as globalised trade. The demand for nearly all coppice products declined rapidly from around the middle of the nineteenth century: railways brought cheap coal to all corners of the nation, coke replaced charcoal in many industries, cheap glazed pottery (made using coke) replaced wooden plates and bowls, metal wire became available for fencing livestock, and iron ships replaced wooden ones.

Demand for specialist items like tanbark continued, albeit inconsistently, along with the ongoing need for things like clogs, bobbins, hop-poles and parts for cart-making, and since the eighteenth century these had been increasingly produced from modified, monocultural coppicewoods whose products could be transported for sale or use in factories. Some coppicewoods were cleared to make way for farmland as the nineteenth century progressed, although their eradication was slowed by an agricultural depression. The same depression also hastened the planting of some fields with conifer under the influence of scientific forestry, a movement that had begun in eighteenth-century Germany and had – relatively slowly – started to influence British landowners. Overall, though, the rise of fossil fuels and their associated materials, along with the importation of timber, saw Britain's woodland cover fall to an all-time low of around 5 per cent as the twentieth century began.

Britain was obviously, by this point, a major colonial power with a global empire and had, since around 1750, been the biggest trading nation on the planet. Timber was one of the most important products exported from British colonies in Canada and America (and subsequently from the United States) and nearly all ships' masts and spars for British naval dockyards were of tall and strong eastern white pine from as early as the seventeenth century. Similarly, planked 'deal' – softwood – was imported in huge quantities from the Baltic states: in 1805 1,000 ships crossed the North Sea carrying timber, a result of the region's seemingly endless forests and cheap labour as well as the Baltic sawmillers' adoption of state-of-the-art mechanical saws.[3]

The German naval blockade during the First World War radically reduced these imports just as demand for both timber and wood increased. The pivotal wartime need for pit props, to support the tunnels dug for coal mining, is well known and was essential to ensuring an adequate supply of coal to fuel power stations, industry and the navy: before the war, all the pit props used in Scottish coal mines were imported from the Baltic, along with those for many of the mines in the north-east of England. Collieries were the biggest consumers of timber in both wars.[4]

More charcoal was also needed for gunpowder, and much military equipment still relied on wood, from parts for weapons to carts and trucks. Clive Mayhew, an arboriculturist who has studied in detail the use of wood in the wars, illustrates the massive requirement for boards simply for packing cases using photographs of docks and marshalling yards stacked with vertiginous piles of munitions, food and all the other equipment that needed to be shipped to the Continent. Many of the other existing needs of peacetime society for imported timber also suddenly had to be sourced from British woodland.

Some of the plantations that had been created under the developing idea of scientific forestry – crops of relatively fast-growing, often non-native trees grown in monocultures – were beginning to mature, but not enough to meet demand, so attention turned to abandoned coppicewoods. Some saw their standard trees graded out for timber, while others were felled entirely if the underwood could be charcoaled or was otherwise useful. Many western oakwoods were cut and the bark stripped for tanneries that were busy fulfilling the demand for leather boots, jerkins and horse tack. By the end of the war 182,000 hectares of woodland had been harvested and imports reduced to less than half of the wood and timber used.[5]

This recent contribution – and the potential future role – of native woods was ignored in the government's response to the timber crisis in the years immediately after the First World War. The Forestry Commission was created to ensure that Britain would always have a strategic reserve of timber, focussed on the provision of pit props, and was given the power to acquire land for plantations.

With coppice already considered an outdated tradition, a 'scientific' approach to forestry was assumed to be more effective, but coppice poles of many of the stronger species could have provided a range of uses even more quickly than planted larch or spruce – although it's often said that miners preferred softwood for pit props because it gave a warning by creaking before a collapse. Having become accustomed to using imported softwoods, it's understandable that the industries that used timber now wanted the same thing from domestic sources. The decentralised and disorganised nature of the remaining coppice industry also played a part, its myriad small, competing firms being unable to work together to present a viable alternative. Britain's woodland cover began to grow again, but the majority of new woodland planted was of non-native conifer species – a situation that didn't change until as recently as the winter of 2021.

The Second World War followed the First too quickly for the Forestry Commission's remit to prove successful, as its new plantations were still mostly immature, so the old woods were cut again. The situation was even more dire this time: the threat from U-boats to shipping was greater, and the woods of parts of France and the Baltic states weren't available to the Allies in the way they had been in the first war. Some of the uses for wood had changed: chemical processes for tanning leather meant oak bark was in lower demand, for instance. Swathes of oak coppice in Britain's temperate rainforest were cut, however, for charcoal that contributed again to gunpowder and explosives, to the production of aluminium, and for the filters in gas masks, among other things. Birch famously went into the 'Wooden Wonder' Mosquito fighter-bomber, along with imported balsa, but its twigs were used in larger quantities to descale molten metal being formed into armoured plate.

Other species of timber and wood were used for the creation of barracks and temporary housing, railways sleepers, dockyard scaffolding, telegraph and tent poles, moulds for concrete pillboxes, temporary roading, and innumerable other purposes on top of the usual civilian uses that had again, in the previous decade, been provided by imports. As in the first war, wood for packing crates was needed in enormous quantities.

With male woodcutters conscripted into the armed forces, the Women's Timber Corps saw 15,000 'Lumberjills' between the ages of seventeen and twenty-four sign up to learn their 'four pillars of forestry': felling, hauling, sawmilling and timber measurement. Canadian loggers brought the latest technology in the form of bulldozers to make tracks and haul timber from the woods, while Italian and German prisoners of war were also set to work. Two hundred thousand hectares of woodland were harvested, which included another cut of some of the woods felled in the first war.

At Spring Park, as in so many woods in the south-east of England, some of the physical traces of the Second World War are so fresh and so integral to the way people use the place every day that it's sometimes hard to think of them as archaeology, although they undoubtedly are. The grid pattern of rides – the main tracks through the wood – is surfaced with old red bricks that are, by reputation, rubble bulldozed from buildings destroyed in the Blitz. The tracks were laid by Canadian soldiers so they could drive their heavy trucks into the wood when they camped there for months in the build-up to D-Day, and I feel a worker's affinity with these young men – perhaps scared and homesick, maybe still excited by their adventure – when I observe the varying quality of the tracks. Sometimes the bricks are laid perfectly, like cobbles, presumably at the behest of a diligent officer, while in other places they've been dumped into uneven piles in a rush, I assume, before a lunch or fag break. Elsewhere among the trees along the length of the whole ridge of gravel that runs from Spring Park to Croydon there are circular pits and the occasional concrete plinth, obscured by fallen leaves, where anti-aircraft guns and searchlights were placed.

The remaining overstood coppice stools at Spring Park, like those in that corner of lime and chestnut I explored with Hugh and Jonathan, are also a direct remnant of its wartime use as all the evidence points to the majority of the wood having been felled some time during that period, but before the Canadians arrived (with the slightly baffling exception of the oak standards).

Clive Mayhew showed me that the wholesale felling of most of the chestnut coppice in Kent, Surrey and Sussex was directly related to the planned landings in Normandy and that much of this material would have been used in the invasion: it was cleft and made into temporary tracking to aid heavy vehicles over soft ground, and fascines for filling trenches and tank traps. The fascines looked just like the rolls of chestnut paling you can buy at a farm supply store today, and were made using the same wiring machines. At Spring Park, the soldiers would have pitched their tents between coppice stools that were shooting back into bushy growth; they were presumably tall enough, by 1944, to provide a measure of camouflage from reconnaissance planes.

Oliver Rackham points out that, while the scale of felling in ancient woodland during the wars was extraordinary, it didn't affect all woods, so plenty of ancient woodland has indeed been neglected since the Victorian era, some since the First World War, and the majority of the rest since the end of the Second. Immediately after the war, the spirit of technological progress and a philosophical commitment to scientific forestry – most significantly within the Forestry Commission – saw the young coppice regrowth of the felled ancient woods dismissed as 'scrub' in its 1947 woodland census, a categorisation that influenced attitudes towards old woods for decades.

Forestry was perceived to produce better timber more quickly and efficiently by the planting of monocultures of fast-growing non-native species like larch, Douglas fir and – particularly – Sitka spruce. The growth of uniform plantations, as well as their likely income, could be forecast more easily, and, not only could their harvest be mechanised, it also required less skill and fine judgement when compared with the myriad nuanced decisions historically made by the woodmen in native woods. Sawmillers, too, preferred timber from plantations, which, being consistent and straight, reduced the amount of time and work needed in adjusting machinery to produce identical end products. And, as we've seen, many of the industries that used timber had by this point adapted to using imported softwood.

More than a third of Britain's ancient woodland was subsequently felled between 1945 and 1975, the stools killed with pesticides or torn out by machines before being replaced with conifers, the planting of which was incentivised by grants and tax breaks. I was interested to discover, however, a 1956 Forestry Commission bulletin called *The Utilisation of Hazel Coppice*, intended to help woodland owners find new economic uses for their neglected traditional woods. It reminded me, again, of George Peterken's advice that there 'isn't a strong plotline' in the history of ancient woods. Even in the mid-Fifties, the most important remaining markets for hazel were reported to be packing crates for pottery, hurdles, thatching materials and beanpoles, and it was noted that the markets were restricted mainly by the lack of coppice-workers who retained the skills to convert their crop on site. Hurdles for garden use were noted as the market with the greatest potential for growth, if more skilled hurdle-makers could be trained.[6]

With rationing still in effect, farmers were encouraged to continue to maximise food production, and they similarly looked to the 'reclamation' of so-called derelict woodland that was assumed to be not only unproductive but actively wasteful in its occupation of land that could be used for crops or livestock. On top of this, a common belief developed that woodland harboured pests that would spill out to affect the rest of the farm.[7]

As agriculture became more mechanised and each farm more specialised in fewer products, farmers also lost some of their general skills of countryside management. From being a valued part of the business of many farms and estates before the first war, woods now became alien places; a new generation of farmers had no understanding of how to manage them, and no reason to learn. The clearance of ancient woodland for agriculture picked up pace and between the war and the late 1970s 50,000 hectares – 10 per cent – of ancient woodland was completely cleared.

Despite the ultimate abandonment of many ancient woods being delayed by the emergency felling of the two world wars, a course was set during the Locust Years between 1945 and the 1970s that would have seen the eradication of Britain's ancient woodland if not for the intervention of Oliver Rackham, George Peterken and their band of free-thinking colleagues. Beyond the clearance of ancient woodland and its 'coniferisation', existential threats that were – mostly – ended by the shift in forestry culture Rackham and co. inspired in the 1980s, the threat from a new dark age remains. The abandonment of as much as 90 per cent of the remaining ancient woodland means that the host of species that rely on the conditions created by light, warmth and a varied structure continue to decline dramatically even in the 'saved' woods. Forest Research, the scientific arm of the Forestry Commission, estimates that only 7 per cent of England's native woodland is in 'favourable' ecological condition, based on their monitoring of a range of features known to influence its use by wildlife – with the worst performing aspects being the lack of both structural diversity and deadwood, along with overgrazing by deer and livestock.[8]

Butterflies form the other iconic group of species whose fate has been severely affected by the decline of traditional management, but it is by observing woodland butterflies that I have seen the most tangible signs of hope. Woodland is home to more of Britain's fifty-seven resident species of butterfly than any other habitat, but dark and shady woods won't cut it. Butterflies rely on the warmth of the sun to fly, while their caterpillars often need specific wildflowers to feed on; only two species can cope with darker woods, the speckled wood and white admiral, and even they can't tolerate complete shade. A number of butterfly species have, like the nightingale, become emblematic of the losses in wildlife caused by the end of coppicing. The Duke of Burgundy, chequered skipper and five fritillaries (the pearl-bordered and small pearl-bordered, heath, high brown and silver-washed fritillary) are mostly restricted to woodland and, although they had already declined dramatically when consistent monitoring began in the 1970s, were much more widespread then than they are now.

Ecologists have been tracking populations of butterflies using the UK Butterfly Monitoring Scheme since 1976 and, between the start of the scheme and 2019, the abundance of woodland specialists fell by one- to two-thirds, depending on species, while their distribution – the number of sites where they've been observed – has similarly shrunk.[9] Their decline, however, is widely held to have been even more dramatic since the 1950s based on previous records simply of their presence, which show them disappearing from many woods – and indeed whole counties – before the consistent survey work began.

The small pearl-bordered fritillary is perhaps the species most closely associated with coppicing among professional ecologists, its reliance over recent centuries on traditional woodland management having been demonstrated by various detailed studies. The 'SPB', like the other fritillaries, has a delicate and sophisticated pattern of black dots and lines on its rich orange upper wings, with tattoo-like chevrons marking their edges, but the underwing lends the species its name, with its series of faintly iridescent triangles. In the occasional woods where it continues to thrive it flits and glides close to the ground along sunny rides, stopping at thickets of common plants like thistles and brambles for nectar. The caterpillars, though, feed exclusively on violets growing in warm, sheltered places, and the adults prefer to lay their eggs on flowers among low, shrubby growth; the spring of coppice two to five years after cutting has been found to best support the species. The cessation of coppicing is thought to be the cause of its complete disappearance from the Midlands and East Anglia some time in the post-war era, and its extinction from Kent and Surrey even in the twenty-first century. Between 1976 and 2019 it decreased by 66 per cent in abundance and 71 per cent in distribution;[10] not only have its numbers fallen

Gavin McCourt making hurdles at Leith Hill, Surrey.
Until recently, half a dozen hurdle-makers made a living here. Gavin is one of the last.

The 'spring' of lime coppice at Shrawley Wood, Worcestershire,
accompanied by foxgloves in its first summer after cutting.

(*Above*) The lapsed oak pollards of Ashtead Common.

(*Right*) Jamie Simpson pruning one of Ashtead's ancient oaks.

Resurgent scrub surrounds a mature oak at Knepp. In the foreground, the work of pigs creates the conditions for more scrub to develop; in the distance, storks nest in a treetop.

Scrub and wildflowers between the pollards in a healthy, restored area of wood pasture at Ashtead Common – following clearance of dense holly and birch.

(*Above*) Highland cattle graze among the trees at Dolmelynllyn, North Wales. (Credit: FSC/Ben Porter)

(*Right*) A broken small-leaved lime springs back to life as a natural coppice stool.

Me on the big lime by the charcoal hearth in Dodgson Wood. (Credit: Edward Parker)

A willow warbler, one of the woodland birds in decline due to the lack of scrub and thickets of young trees. (Credit: Getty Images/Morgan Stephenson)

Typical symptoms of ash dieback in the Peak District.

Daniel Brown, Dave, Rita and Eddie hauling a chunky piece of sweet chestnut.

A beaver-gnawed stump I stumbled across in Tayside, Scotland.

The decaying wood and rich fungal ecology of the primeval forest of Białowieża, Poland. (Credit: Zuza Featherstone)

The giant lime in Dodgson Wood in 2015, after the final section had failed.

by two-thirds, but it has become locally extinct in two out of every three woods where it was previously observed.

Of those seven rare and fast-declining butterflies of coppice-woods, only the silver-washed fritillary has been seen at Spring Park in living memory, and I occasionally saw these orange giants gliding confidently around the sunnier rides. Of the woodland butterflies, the silver-washed 'frit' can best cope with a mature canopy, as long as it's not too dense and plenty of sun's getting through. More common, generalist butterflies can still help understand the health of the woodland ecosystem and the rangers have been monitoring them at Spring Park and Ashtead Common since 2002. I used the subsequent dataset and compared it with the specific types of woodland habitat along the survey routes for the thesis for my master's degree in forest ecology – which I somehow completed in the evenings while working in London and the Lakes.

Surveying for the UK Butterfly Monitoring Scheme involves walking a consistent 'transect' – a set route around the site – every week during the summer and recording every butterfly observed in an imaginary box around and just ahead of you. It can be difficult as a ranger to make time for wildlife surveys, as there's always a more urgent job that needs doing with a tangible outcome, a gate that needs mending or a path strimming. But not only are the long-term results of surveys like this incredibly important for informing our work locally and understanding the national picture of biodiversity decline, I also always found walking the transect invaluable in reconnecting me to the sites I cared for and refreshing my motivation. It was a meditative experience, once I knew the route by heart, to steadily walk Spring Park's paths focussed solely on spotting and identifying the flittering, spiralling insects that crossed my path in the dusty summer wood.

Each time I walked the transect speckled wood butterflies were easily the most abundant species, circling in pairs and threes through shafts of sunlight, their brown and white markings camouflaging them when they rested on bramble leaves in the dappled shade. And, before I even studied the numbers, I knew without doubt that I saw more butterflies of a wider range of species – white, yellow and blue, and gleaming black and red – in the sunnier places: I recorded brimstones, large and small whites, orange tips, holly blues, skippers, white and red admirals, peacocks and more. Nothing, perhaps, that anyone who notices butterflies would think is particularly unusual, but lovely to see, and the specific locations where I saw them told a powerful story about the way that even the more common species use woodland. Although most butterflies are struggling more widely, the key to these generalists' relative success in comparison to the woodland species is that their caterpillars are either less picky in their requirement for a specific food plant, or rely on a much more common plant that doesn't require the particular conditions associated with coppicing; the peacock and small tortoiseshell butterflies, for instance, lay their eggs on nettles, while the holly blue will use a range of shrub species.

Having collated both my few seasons of data and that from preceding years, I broke the transect routes down by habitat, discerning between sections of mature woodland, woodland edge, coppice coupes, scrub and rides; I also measured the shade cast by each section's canopy to confirm that these categories corresponded with increasing levels of light. Comparing the numbers and species of butterflies with the types of places they'd been seen at both Spring Park and Ashtead confirmed my observations and the more intensive studies of the real butterfly experts: the lowest numbers of both species and individuals were in the darkest areas of mature trees, while the surveyors had seen many more butterflies on the edges of the woods, along sunny open rides and, particularly, among the scattered scrub of Ashtead Common.

At Spring Park, I was able to measure a slight increase in the numbers of butterflies spotted in coppice coupes that had been restored and recently cut, but the difference wasn't as dramatic as the books suggested it should be. Upon examining the levels of shade in the coupes more closely, the reason was clear: our cautious felling of canopy trees even after the understorey had been cut meant they still weren't sunny enough to meet their full potential as butterfly habitat, adding further impetus to our decision to increase its rate. Sure enough, more recent butterfly surveys seem to demonstrate an increase in the numbers of butterflies using freshly cut coppice where more standards have been removed.

Elsewhere in the country, a few conservation-focussed coppice restoration projects are proving successful in sustaining remnant populations of rare woodland butterflies. Blean Woods National Nature Reserve, not too far from Spring Park in Kent, is the flagship example, where sophisticated coppice regimes linked by rides and glades have been cut since the 1980s specifically to support its population of heath fritillary, which relies on common cow wheat that appears in the first couple of years after coppicing. The heath fritillary was traditionally called the 'woodman's follower', so important is the coppice regime to its survival. In Hawkcombe Woods on Exmoor, fifty of these dusky butterflies were reintroduced in 2014 following a programme of coppice restoration in this oak-dominated temperate rainforest, and they quickly increased in number to form a stable ongoing population. Around Witherslack, meanwhile, not far south of Dodgson Wood and in the same area of oakwoods sustained by the historic iron industry, there's a northern stronghold of both coppice culture and, not coincidentally, woodland specialist butterflies, where well-designed coppice management is keeping pearl-bordered and high brown fritillary populations alive.

The charity Butterfly Conservation runs the UK Butterfly Monitoring Scheme and periodically publishes reports on its results. The 2022 report makes for gloomy reading in that 80 per cent of British butterfly species are in decline when compared to the 1970s; more hopefully, they list lots of examples where targeted conservation work – like the projects in Witherslack and on Exmoor – make a real difference in boosting numbers. The report notes, though, that the resources available to expand this sort of work are 'woefully inadequate'.[11]

Like many of the plant species of ancient woods, the woodland specialist butterflies struggle to colonise new sites even if the conditions are right as they simply don't fly far from their home territories, so it's unlikely that any of the smaller fritillaries will make it back to Spring Park any time in the foreseeable future. It's a shame, to put it mildly, but the restoration of good woodland management can still play an important part in increasing the abundance of more common butterflies, making our conservation efforts at Spring Park worthwhile. And, as they're indicators of the health of wider insect populations, we know that, if butterflies are doing better in the restored coppice coupes and rides, lots of other species will be too.

Mike, Jamie and I woke at four and met back under the dead oak to watch and listen for any bats returning to roost under its bark, Jamie ready with his climbing gear. The woods are even more silent in that hour later in the summer, the dawn chorus months past, and we were groggy and subdued too. There were fewer pipistrelles and little of the excitement of the previous evening, and we stood quietly enjoying the coming of the light. No bats flew toward the oak, and, as soon as the sun appeared through the understorey and we'd gone a few minutes without any action, Mike told us we might as well get on with it. Jamie climbed up and knocked the big limbs off the path-side of the oak with his chainsaw, letting them tear out to leave stubs that he dressed slightly, trying to leave a natural-looking fracture rather than a harsh, circular wound. We cleared the cut branches into the bramble and were at the cafe eating egg, chips and beans before seven.

The need of many bat species, pipistrelles included, for both open areas within woodland and roost sites like flaking bark and rot holes in old trees begins to demonstrate the challenge facing today's woodlanders. While we know that the end of coppicing has played a big role in the precipitous declines of many species in our remaining ancient woods, the solution isn't as simple as just restoring the coppice regime everywhere – even if that were possible. We also need to factor in the 'old-growth' areas where trees can mature, die and decay undisturbed for decades and centuries, providing roost features for bats as well as birds like woodpeckers and tits; about a third of woodland birds nest in holes in old trees.[12] They also form essential habitat for the thousands of saproxylic species like those in Ashtead's oaks – although there are a whole different lot of beetles, wasps and flies that rely on the decaying wood of dense old woods, as opposed to open-grown ancient trees.

While many ancient woods have been neglected in terms of the active management they need to let in light and warmth, there has also over recent decades been a forestry-inspired trend for tidying woodland, with woodland managers removing fallen trees and standing deadwood almost as a point of pride. The habitat provided by decaying wood is now often missing, along with the lack of structural diversity, although most people caring for woods for nature are at least now aware of the former issue. At Spring Park, around a quarter of the wood was purposely left undisturbed when Baz and the other rangers restored the coppice regime, these dark corners slowly continuing to tangle and ripen as mature trees collapse and are left to crumble and collapse. Jamie, likewise, has designated areas of Forge Wood 'minimum intervention', in the phrase of the management plan, and has strategically located them around badger setts and sensitive archaeology that he needs to protect anyway.

Although as a ranger I was closely involved with coppice restoration – and it has a particular appeal, with its long history and associated crafts – coppicing isn't the only way to achieve the shifting areas of light and warmth that our declining woodland wildlife needs, and other methods might help provide more essential old-growth habitat. In some places, a novel approach to forestry has been applied to ancient woodland, and might be even better for wildlife while also providing more useful timber for modern needs.

'Continuous cover forestry' or 'CCF' is an alternative to the conventional post-war forestry system of 'clear-fell and restock' that sees uniform blocks of single species planted in rows before being cut in big patches – at least the size of a couple of football pitches – then replanted; this is the simplest form of scientific forestry. It's efficient and easy to manage and produces a quick, consistent crop of timber.

In Switzerland and other parts of central and western Europe, however, the shortcomings of clear-felling were recognised as early as the beginning of the twentieth century as the big patches of cleared ground became prone to landslides without the stabilising effect of trees. Continuous cover forestry was developed in response and is now common on the Continent, and has started to be introduced into British woods over the last twenty-five years – although the majority of commercial forestry is still managed under the clear-fell system. Under CCF trees are thinned more frequently but in smaller numbers, either individually or in small groups, in a process termed 'selective felling'. The resulting, ongoing structure eventually consists of stems of all sizes, with light levels carefully managed to provide suitable conditions for seedlings, saplings and maturing trees, as well as understorey species.

The most vocal advocates of CCF in the forestry sector are understandably focussed on the economic benefits – it requires much lower inputs in terms of planting and ground preparation, and can provide a better and more regular stream of income from higher-quality timber. Stands are more resilient to pests, diseases and other threats due to their species and structural diversity, and its use also sees a much lower impact on the precious woodland soils and avoids the pollution of watercourses associated with clear-felling. Most relevant for ancient woodland, though, is the fact that CCF can also be much better for woodland wildlife than either clear-felling or, at the opposite end of the spectrum, non-intervention or neglect, and is sometimes actually called 'close to nature forestry'; the gaps and glades can be designed to allow ground flora, birds, butterflies and the whole range of woodland wildlife to thrive, while retaining areas of old-growth habitat throughout.

At its best when applied in ancient woodland, selective felling can be used to replicate the abundant habitat features of the wildwood even more accurately than by coppicing, while the closely controlled approach can simultaneously produce timber of exceptionally high quality. The National Trust's characterful forestry adviser in the post-war era, John Workman, was an early and passionate advocate for CCF. He reminded visitors to his carefully tended wood at Ebworth in Gloucestershire that it was part of a National Nature Reserve and said that 'the same insects live on good trees as bad, so you might as well have good trees and make some money'.[13]

One of the many students who Workman inspired, Andy Poore, has gone on to be a prominent voice of CCF and has developed the practice on estates across southern England, including at the Rushmore Estate on the border of Dorset and Wiltshire. While not alone in adopting a selective approach to felling, those involved at Rushmore are almost unique in having studied the impact of CCF on the birds, moths and bats in the estate's ancient woodland. Ten of twenty woodland birds monitored are doing better in the CCF areas, including blue tits, wrens and blackcaps, although five species continue to thrive more in areas of coppice, like the migrant willow and garden warblers, underlining the need for a variety of approaches. Andy told me that the complex CCF habitat at Rushmore supports particularly high populations of woodland species that are suffering some of the worst declines elsewhere, like marsh tit and Barbastelle bat, while other bats and invertebrates have also responded positively. All the woodland that's under active management – that is, where trees are sometimes felled – shows higher levels of use by birds, moths and bats when compared to neglected areas.[14]

This approach to managing ancient woodland as 'high forest' seems likely to play an important role in the future of our old woods; it is, in some places, more attractive in economic terms to landowners and can be tended by foresters, requiring less specialism in the form of coppice-workers while producing wood more suited to modern uses. Most importantly for me, it has the potential, with thought, to recreate even more closely the natural woodland in which all species evolved, admitting light and maintaining old-growth features to create the disordered complexity to which wildlife responds so well.

The closest I've come to experiencing the amazing synergy between traditional woodland management and a flourishing, more natural ecosystem is when I visited the Maramureş region of rural northern Romania to meet a family who my parents had befriended on one of their youthful adventures in the 1970s. Sat at dusk beneath a spreading walnut tree outside their log-built summer kitchen, we glimpsed something of the spectacle that must have graced British woods until the twentieth century. The air was alive with bats, zipping and looping around our heads and between the apple trees and round-wood buildings of the farm. I have no idea which species they were, although I suspect plenty of them were pipistrelles; but it didn't really matter. The sheer quantity was extraordinary, and also illustrated what must have been a booming population of the insects on which the bats feed.

Maramureş and its surrounding provinces represent some of the most intact remnants of a pre-industrial agricultural system in Europe. The area holds a particular resonance for British conservationists, as the geology and ecosystems are reasonably similar to somewhere like Derbyshire's White Peak, so we find lots of familiar species but in unusual and moving abundance, not to mention once-common species that have been driven to extinction in Britain, like the red-backed shrike. I walked through informal meadows that would have been the jewel in the crown of an English nature reserve, our friends – and everyone else – in a constant process of casual, almost absent-minded scything. The distinction between fields, as we know them, and woodland glades was entirely missing; with livestock still controlled by shepherding across what we would call common land, the landscape was a loose patchwork of woodland, scrub and open-grown trees among tapestries of bright flowers spread as far as we could drive in a day. It was wood pasture on a genuinely landscape scale, and filled with those species that are merely clinging on in Britain's compartmentalised countryside and degraded habitats.

But it is far from wild, and remains unmistakeably a wood culture, with everything built from the products of a system much like coppice in the denser bits of woodland: from tools, houses and wood-fuelled summer kitchens to the giant decorative gates of the smallholdings, the atmospheric wooden churches that form a UNESCO World Heritage Site, and the carts that are still a common mode of transport. Wood produced locally remains an essential component of everyday life.

It's an enchanting place to visit, but the reality of life there is hard. Many of the remaining farmers only eke out a subsistence lifestyle, while younger people understandably look to cities, like the nearby digital business hub of Cluj-Napoca, with their more modern forms of employment. As in Britain over previous centuries, the ecology of the region is beginning to suffer from abandonment in some places, and the intensification of agriculture in others.

A new wood culture in Britain can't seek to recreate what came before, nor to replicate the remnants of something older, which we still find in places like Maramureș; we have to find ways to make our ancient woodland relevant to modern society, and to our current and future needs. But most pressing among those needs is that of helping nature recover, and in this we can take inspiration both from the history of coppicing and from more novel forms of silviculture – as well as by applying our increasing understanding of the pre-agricultural wildwood. The contrast between what I experienced in Romania and the few bats we saw at Spring Park – exciting and inspiring as they were, too – brought home to me what we've lost on our nature-depleted island; but they also illustrate what we stand to gain by rejuvenating our woods with an influx of light and warmth, and by allowing their life to spread back into healthier landscapes around them.

CHAPTER EIGHT

Darkness and Light

Ash dieback and other threats to ancient woodland

TREES BLANKET THE other slope of the valley as it rises tall and steep, curved around in front of me like a vertiginous amphitheatre. It's an Imax of woodland that entirely fills my vision, stretching in both directions until the dale twists and turns and the wood disappears around the corners. Stood on the boot-smoothed limestone promontory of Lover's Leap, high above the river, I realise that I'm watching Dovedale Wood change on an unusually human timescale; the wood is markedly different from how I first knew it only a decade ago.

In any wood in summer, the word 'green' becomes completely inadequate. I pick out the dark shades of yew clinging on around rocky outcrops and solid, military blocks of beech, planted here in the Victorian era to further 'beautify' the valley. But the view is dominated by a shifting canopy that is richer and brighter, as well as lighter and looser in texture. It seems to flow down the hillside to break around the limestone spires known as the

Twelve Apostles, where a heron is perched incongruously on one of the rough pillars.

Five years before, in 2016, I'd stood in the same spot with Joe Alsop, the site manager with Natural England for Lathkill Dale, a few miles to the north. He'd shown me the first subtle signs of ash dieback in the tips of saplings beside the path; the canopy opposite, though, had looked full and reassuring in the face of the disease that we feared would radically alter these woods. Now, my eye is drawn unavoidably to the grey among the green wall: there are ash trunks visible through the thinning canopy, while a couple of stands of trees – perhaps a dozen in each – are either completely dead or wear only a feathery smattering of leaves along their trunks. Skeletal branches stick up beyond this last gasp of life like upside-down lightning bolts. When I get in among the trees, the branch-ends are stubby and ragged, the twigs and buds long gone, and brittle blobs of King Alfred's cakes fungus dot stems pockmarked by woodpeckers. I put in hundreds of gruelling hours lugging my chainsaw and planting spade up and down those slopes, trying to mitigate the potential impact of dieback; the rest of the ranger team – Kate, Andy, Mark and the others – put in even more graft. We had no idea this would happen so quickly.

New tree diseases, often unwittingly imported by globalised trade, pose a major threat to our ancient woods, but they're far from the only one. Working in Dovedale, at the southern edge of the Peak District, I was forced to consider how far we might need to intervene to ensure the resilience of our ancient woods in the face of disease and a raft of other, compounding pressures, which range from too many deer eating their future to invasive plants that swamp woodland wildflowers. Those of us involved in looking after the White Peak's ash-woods had to learn to balance a philosophical acceptance of their inherent dynamism with the urgent need to protect their unique characteristics; a particular challenge here with the woods displaying far fewer signs of human use than most, and a myth having developed that this means they're 'wild'. Even as I began to understand this constantly evolving ancient landscape, I experienced a process akin to grief as I watched it change with astonishing and unnatural speed.

I knew when I moved to the Peak District from the Lakes that my time here, in the ash-dominated landscape of the White Peak, would be defined by *Hymenoscyphus fraxineus*: ash dieback. The disease made headlines in 2012 when infected saplings imported from mainland Europe created outbreaks across the country – although it seems likely it had arrived here naturally before then, the English Channel being no obstacle to microscopic, wind-borne fungal spores. I'd seen its impact on visits back to the south-east, where whole ash-woods were starting to look thin and ragged, but there had been no signs of it in the historic ash pollards on my patch in the Lakes. The woods of Dovedale and the surrounding dales looked similarly unaffected but, with ash being easily their most common tree, it was obvious that its inevitable arrival would cause dramatic changes.

The White Peak has sometimes been called a landscape in reverse, formed as it is of a raised plateau of limestone now covered by dairy farms across its gently rolling fields. Unlike most British uplands, where things get wilder as you gain altitude, the rugged, nature-rich dales are cut into the plateau so you drop from the silage fields into deep valleys of species-rich grassland and ancient ash-woods among towering, gleaming crags. The slopes are too steep to have ever been ploughed or had fertiliser spread on them, so they remain home to some of the richest habitats in Britain, the calcareous substrate supporting a wide range of wildflowers, butterflies and other species in both the grassland and the woods. Dovedale is the longest and most dramatic of the dales, winding steep-sided for almost ten miles and punctuated by looming, characterful rock formations.

Dovedale Wood, in turn, is the biggest native woodland in the White Peak at more than a hundred hectares; like Dodgson, this is a big wood by any British standard. All the other ancient woods nearby are similarly confined to the incised dales and the trees are, we think, around 80 per cent ash, while the species' dominance of the area is even more evident to the visitor in the field trees that line the roads and dry-stone walls up on the plateau. Ash, with its light grey bark and bright compound leaves, is so integral to the character of the landscape that I feel sure it contributes to the name the 'White Peak', the trees complementing the limestone outcrops and walls to which the label is usually attributed. On clear-skied summer days the delicate edges of their crowns seem to have a softer relationship with the sky than other, denser trees, the wispy rows of leaflets forming a shifting edge that pales and bleeds to form a new colour between the green and blue like misprinted news-sheet.

Like Britain's other ancient woods, this landscape of ash is the result of both natural and human history. The limestone geology itself suits ash, which is better able to tolerate its alkaline soils than oak, while other canopy species like beech and hornbeam didn't naturally make it back this far north after the last Ice Age; ash dieback is a reminder that we don't actually have many big species of tree native to Britain, and the human modification of our woods has favoured even fewer of them. The ecological influence of the underlying rock is exacerbated in the dales, where ash is better able to deal with both the thin, nutrient-poor soils and the unstable terrain of the screes, its roots quicker to react to the shifting surface where lots of tree species simply can't cling on.

Quarrying, lead mining and changes in grazing influenced many of the woods in the dales more than their management for timber and underwood, but the impact of these activities was to drive out species like both small- and large-leaved lime that would have formed a bigger natural component of the canopy but aren't able to recolonise the areas from which they've been lost due to their reliance over recent centuries on vegetative reproduction. Ash, on the other hand – as any gardener knows – is a 'pioneer' species that produces lots of seed and spreads rapidly into new ground, hence its current dominance and the vulnerability of landscapes like this to ash dieback.

The fungal disease originated in East Asia alongside the Manchurian and Chinese ash, sister species of European ash, as the tree native to Britain is properly known. Having evolved together over millennia, what we know as ash dieback only affects those other species of ash as a leaf blight – a minor pest that they've learned to live with. It was first observed in Europe in Poland in 1992, having hitched a lift with imported trees or other cargo, and has been making its way steadily across the Continent since then. (There's a chance that it could, at some point in the future, have made this transcontinental journey naturally, but, given the prevailing winds and the mountain ranges in its path, it's considered unlikely.) European ash has no evolutionary adaptation to the novel pathogen so, instead of being held at bay in the leaves, the fungus grows down inside the tree through its vascular system. With the vital water- and carbohydrate-transporting cells blocked, the crown starts to die back, characteristically from the outside in as the tips of the twigs and branches are affected first; tiny fruiting bodies on the dead leaves produce more spores, which sail off on the wind to infect new trees. No treatment or cure is possible, so all that woodlanders can do is manage the consequences.

Joe, the site manager who tuned me in to the first signs of infection in Dovedale, had previously travelled around Europe learning about ash dieback and told me that its impact on individual trees was very variable, perhaps because ash reproduces in such great numbers by seed, so there's a lot of genetic variation, or perhaps because of differences in soil type and climate; nobody's quite sure, although we're learning more each year. The worst-hit places on the Continent have seen more than 90 per cent of their trees affected, while elsewhere its effects haven't been so pronounced. Sometimes the fungus colonises and kills just a tree's most recent growth in any one summer, which is what Joe had shown me on the saplings at Lover's Leap: there was obvious browning of the dead 'leader', the topmost twig, while the branches below arced up to take over in a characteristic trident shape. On bigger trees the crown typically shrinks over a few years as the fungus works its way down, although some trees can die within a year or two of infection if they're already stressed. But others hang on for many years with gradual decline and, particularly in older, open-grown ash, there are signs that some are springing back and developing healthy new crowns rather than being doomed to a terminal diagnosis.

Once Joe had shown me the first signs of dieback in the saplings, I began to see that trident shape everywhere I went in the White Peak. Although it was concerning, it was also galvanising that the threat was no longer theoretical but tangible and imminent. With a new sense of urgency, we began making a plan to try, in some small way, to mitigate the impact of the disease in Dovedale Wood.

The first job, when you're writing a plan for the management of a wood, is to do your research, and that starts with getting to know the ground. The other rangers and I spent that first spring and summer surveying the Trust's woods in the White Peak to understand their intricacies. It was a ridiculous privilege to spend weeks on end walking them, finding their hidden spots and special species, even more so given how inaccessible these woods are and how few people get the chance to explore them. But it was bittersweet. We would be the last people to experience Dovedale Wood and its counterparts as healthy ash-woods.

Dovedale felt different to nearly every wood I'd worked in before and the meandering surveys around its steep slopes took me through increasing layers of unfamiliarity. Most obvious was the amount of ash, which was like making a new friend after years of knowing oak so intimately, and at first I missed the familiarity of the gnarled, solid trees in their many different forms and settings: from Spring Park's standards to Ashtead's ancient pollards and the rainforests of the Lakes. But soon I came to love the airy feel of the light canopy that seemed constantly in motion, and as spring progressed we began to see the wildflowers for which ancient ash-woods are famous; they're home to some of the most diverse ground flora of any British woodland type.

Among carpets of unshowy and slightly rank-smelling dog's mercury, constant companion of ash on the limestone, there were displays of wood anemone even more dramatic than those at Spring Park, a heart-lifting sea of stars shifting bright across slopes that stretched into the distance. There were violets tucked away among the sward almost everywhere and we constantly had to watch our step to avoid lords and ladies. In the wetter flushes where slight depressions gathered deeper soils, wild garlic filled the understorey with a blinding glow as its little white flowers burst where the spring sun streamed late over the crest of the dale. The track along the river, meanwhile, was lined with deep crimson banks of water avens beneath damp stands of alder, the flowers pendulous and nodding like a crowd of hooded monks.

But despite their sometimes overwhelming spectacle, those species are some of the more common flowers of ancient woodland. There were smaller stands of particularly special flowers that are restricted to the ash-woods, which hold such high value for plants due to the limestone geology, in which lots of wildflowers thrive, but also because of the canopy of ash, which is relatively sparse and lets just the right amount of sunlight to the ground: enough to fuel the flowers all summer, but not so much that bramble and other coarse species take over. We found sweet-scented and ridiculously delicate lily of the valley, mats of woodruff and tiny groves of herb paris around the base of veteran ash trees or tucked among ice-hewn boulders. The flower of herb paris is wispy and inconspicuous but its four identical leaves, arranged into opposing pairs, make it particularly distinctive; it supposedly inspired the symbol of the true lover's knot, one of its other common names, and was in the past used in marriage ceremonies due to the symbolism of its symmetrical form. Like lily of the valley, it grows from a slowly creeping rhizome and each flower produces just one big seed each year, making them both strong indicators of ancient woodland.

Much more conspicuous was mountain currant, a rare plant elsewhere in Britain but here so widespread that we came to curse it as impenetrable, chest-high thickets – not dissimilar to neglected allotment redcurrants – forced us on exhausting detours up- or downslope. A visiting botanist showed me wood barley between the ruts of the track, a rare grass that I'd never have identified alone. Their hunt for mezereon, a small, slightly anonymous shrub, was fruitless. Mezereon is known for its beautiful pinky-purple flowers that appear before the leaves, meaning that to stand a chance of seeing them you really need to know where to look. Although it's been recorded in Dovedale, nobody around then was quite sure where.

But beyond the ash and its special floral ecosystem, I realised, Dovedale Wood was different in another aspect, and pinning it down left me feeling weirdly adrift in a way that was both disorientating and exhilarating. There's only one short track along the river, and in the absence of a path we beat our own routes across screes and up slick, thin soils where none of the usual vague human marks in the landform of the woods emerged into focus: no charcoal hearths, no woodbanks, no cut stumps, no old walls. I had never been in an ancient wood in Britain with so little evidence of people. And in the depths of the wood, on the very steepest slopes, we found sure signs of its timeless continuity: occasional thickets of small-leaved lime that had clearly been wending their way around the dale for centuries or more. More common were vales of yew clinging to crags, with rare rock whitebeams evident as spring progressed and we were able to pick out the bright white undersides of their leaves.

We found caves in the hollow limestone and fern-filled ravines that felt like lost worlds, and learned that these sheltered spots had yielded extraordinary archaeological deposits that spoke of deep time rather than recent history; I was shown by torchlight around labyrinthine caves where the bones of lemmings and ptarmigan had been discovered, while a colleague produced a bear tooth wrapped in tissue paper from their desk drawer. It looked like a giant human molar, and the cream enamel and brown-stained crevices were so bright and fresh it could have been pulled that morning from an ursine jaw, but it had been found beneath the earth floor of a shallow cavern where it had been buried for 14,000 years. Over the river, the 'Dovedale Hoard' of Iron Age coins, uncovered from the detritus in another cave, added to the sense that, for the people who had used the valley, it was a place of refuge or hiding rather than of everyday work.

Our desk research and conversations with colleagues quickly confirmed that there's no record of traditional management like coppicing in Dovedale Wood or many of the other White Peak woods, making them very unusual given the fundamental relationship between most of Britain's ancient woodland and human industry. We know from photographs that some trees were cut along the valley bottom during the Second World War, but this was probably an exception. Other than that, the woods were simply too inaccessible and remote from any conurbations that needed firewood at scale, as well as from industries that required charcoal or wood products.

We also learned that, since the early twentieth century, the lack of traditional management and the presence of the ancient limes had led influential ecologists like Sir Arthur Tansley to describe the White Peak woods as being entirely 'virgin' woodland, what one book called 'the natural and unspoiled covering of the carboniferous limestone'. But, clambering around the crags and sliding down mossy screes, the rangers and I began to read a different story, and as we talked more to Joe it became apparent that a myth had developed around the place. The areas where we typically found more tree species, particularly the timeless tangles of lime and rock-clung yew, were often around the rocky pinnacles or the steepest scree slopes, the less common trees hanging on in pockets of genuine ancient woodland where livestock had, I suspect, rarely found their way even before there were fences. Like Dodgson's ghylls, these could indeed be vanishingly rare fragments of the wildwood that have probably never been cut or disturbed by people, and they're very special.

Large areas surrounding them, though, were monocultures of ash a few decades old that suggested the trees had only recently colonised open grassland; sure enough, Tony Robinson, a retired Natural England officer and more recently a research student, showed us his collection of old postcards of Dovedale, which proved that some of the now-wooded slopes had been grazed bare within the last fifty years. In between the two extremes, there was a dizzying range of intermediate histories to decipher, and in some areas we could read what must have been periods of near-clearance that had left a cohort of mature trees surrounded by a more recent generation of scrub and young ash. The woodland had clearly waxed and waned dramatically over centuries, and even over the past few decades; mainly, here, in response to grazing pressure.

The oldest, very special bits of woodland around the crags formed a series of nuclei from which the trees had expanded each time grazing had been reduced, and back to which the wood had contracted when the value in livestock was higher. Just like at Dodgson, a big fence had been laboriously built around the whole wood in the 1980s to protect it from sheep. So people may not have cut trees in Dovedale, but our ancestors still transformed the valley by the way they grazed it, with ash best suited to rapidly recolonise the ground cleared by nibbling livestock.

We read more of this nuanced story in the ground flora. Although the richest assemblages of wildflowers were typically in the oldest bits of the wood, the fact that supposed ancient woodland indicator species were spread throughout the valley revealed a deeper significance to these indicator plants: lots of them had clearly been able to hang on when the wood was grazed back. Similarly, in the steep grassland above the trees we found patches of wood anemone and celandine among the cowslips and early purple orchids – even bluebells as the soils became less alkaline towards the brow of the hill – suggesting, in simple terms, that they were probably previously wooded. Woodland expert Ian Rotherham has popularised the evocative concept of 'shadow woods', lost woods discernible from their remnant ground flora.

The so-called indicators of ancient woodland are, though, in some places, so much more than that; they are indicators of ancient, undisturbed *landscapes*, signs that the soil has never been ploughed or disturbed even as trees have grown and disappeared and grown again. It's probably a more accurate representation of the history of Dovedale to think of it as a wood pasture landscape with dense areas of woodland, scrub and open grassland having all come and gone over decades and centuries as grazing changed with agricultural depressions or increases in the price of wool; and those shifting habitats have only been compartmentalised and their boundaries made semi-permanent recently. Some ecologists now refer to ancient woodland species like these simply as indicators of undisturbed ground. This doesn't diminish their value in helping confirm that a wood is old and special, it's just that they can also help tell a wider story of how that wood, and the landscape within which it sits, has changed over time. Understanding the wider history of Dovedale's ancient landscape gave us the confidence we needed to intervene in response to the impending arrival of ash dieback, as well as helping us refine our plans to respect this less obvious human story.

Despite Joe having helped me spot the early signs of dieback, it still felt all but impossible to conceive of the flower-filled woods we experienced on those surveys changing. I saw it in some of my colleagues too: as we looked at Joe's photographs of collapsing, skeletal ash-woods in eastern Europe there was a sense of disbelief, even that we were being alarmist, a dismissive sense that *it can't be that bad*. I began to understand our human inability to confront the threat of climate change. We simply haven't evolved to act in the face of future threats that we're required to imagine in order to respond to; humans were too busy, until the last few decades, dealing with immediate, physical problems. Walking and scrambling among those swathes of wildflowers brought the threat of dieback home to me, as well as that hard-won understanding of how depleted the canopy of Dovedale Wood has become, its diversity worn away by grazing until it became this near-monoculture of ash.

Fortunately, it is only a *near*-monoculture, so the main component of the response we ended up planning was also the least controversial. Even away from the heartlands of untouched, ancient habitat, which we intended to leave well alone, we had found other native tree species. There were field maples in apologetic groups under the canopy, or occasionally one soaring alongside the ash and fooling us, on first glance, into thinking it an oak. There was the odd oak itself, in the richer and more stable soil where the slope was less severe. There were banks of hazel and hawthorn scrub, as well as more unusual understorey species that, like ash, do well on the limestone: spindle, guelder rose and dogwood. As spring arrived, we realised that the space beneath the ash canopy was full of wych elm, their bracts glowing a vivid yellowy green throughout the still-bare wood. Inspired by Joe's plans in Lathkill Dale, we decided to fell ash around stands of these other species to give them space and light to seed into, with the hope that it would boost their regeneration and encourage their expansion ahead of the decline of ash; lots of ash saplings would probably spring up too, but the fact that each one would be genetically unique would increase the chances of some of them being resistant to dieback.

More contentious, although also following Joe's lead, was our intention to plant missing tree species back into the areas that are dominated exclusively by ash. Planting trees in ancient woodland is an emotive subject among conservationists, due partly to concerns about the introduction of inappropriate tree species – the coniferisation of the Locust Years casts a long shadow. Even where the species themselves are not at issue, there's a justified desire to preserve the unique genetic heritage of our oldest woods. There has also, however, almost certainly been a prejudice against planting based on a misunderstanding of British woodland history, with the current species composition of our woods often assumed to be entirely natural and the modification and other human influences of recent centuries ignored.

Planting lime is even more challenging due to the species' unusual and evocative role as a very strong indicator of ancient woodland – and because of the passionate, albeit niche, interest in its remnant specimens as Rackham's 'living links to the wildwood'. I have, though, a hand-typed memo from Professor Pigott to the National Trust that was included in the package of old notes that he kindly gave me along with some of his woodland library. Dating from the 1980s, when he was serving on the Trust's advisory council, it articulates how to go about planting lime in such a way that the historical and ecological significance of remnant populations isn't compromised. We followed his guidance in designing planting schemes in Dovedale Wood that would keep our new limes – and other species – well distinct from the ancient specimens, allowing us to seed the future canopy of the wood with a more natural range of the under-represented species that have been nibbled away by sheep, and that should collectively mature to recreate the conditions for the special ground flora to continue to thrive.

The impending arrival of ash dieback was not the only sobering aspect to our surveys of the White Peak woods. Along the top edges, where the trees crested the slope and met the farmland of the plateau at stone walls or barbed wire fences, chest-high nettle-beds showed where fertiliser, spread as powder on the silage fields, had drifted over the boundary and enabled coarse vegetation to romp away at the expense of the more unusual species of woodland wildflower. It's a common side-effect of intensive farming next to woodland and we were grateful that Dovedale is such an unusually big wood; the impact of a 5-metre strip of artificial nutrient enrichment around the edge of a small wood can be catastrophic for its ground flora.

Air pollution creates a similar issue here too, although it's harder for us on the ground to gauge because its effects are longer-term and more difficult to observe. The Peak District has long suffered from being surrounded by some of the north of England's biggest cities. Most dramatically, acid rain caused by pollution from cotton, wool and steel mills in the nineteenth and early twentieth centuries contributed, alongside its management for sheep grazing and grouse shooting, to the complete defoliation of the moor on Kinder Scout and its conversion to a black moonscape of bare, eroding peat, turning this giant store of carbon into a source of emissions (the restoration of Kinder over the last thirty years, by the way, is one of the biggest and most underrated conservation success stories I know of). Although the impact on the moor was extreme, there would certainly have been severe, albeit less obvious, effects on the area's woodland. Today, nitrogen produced by cars and trucks, power stations, industry and domestic heating is deposited on the original British national park at such high rates that its levels in the soil of Peak District meadows continue to increase despite a crop of hay being taken off each year, reducing the value of these special habitats for wildflowers – another powerful indication of the less easily observed impacts that must be happening in the woods.

In other areas, nitrogen pollution also comes from ammonia from intensive livestock production, particularly from the most intensive indoor pig, poultry and beef units. As well as simply encouraging coarse vegetation, with knock-on effects for the invertebrates and other species that rely on particular woodland plants as the base of the food chain, nitrogen can unbalance the soil chemistry, impeding the absorption by woody plants of other nutrients. It also affects the web of mycorrhizal fungi that live in symbiotic relationship with trees and which are vital to their health.[1]

Nitrogen is particularly bad news for lichens and mosses, and part of the reason places like Dodgson Wood remain so rich in these groups is that, as well as their wet and mild climate, the prevailing winds arrive unpolluted from the Atlantic. There are pockets of the Peak District that meet the climatic conditions for temperate rainforest, and Dovedale itself is home to rare ferns and other plants that rely on the humid microclimate of its deep gorge. It's tempting to wonder whether the naked trunks of Peak District trees would once have been clothed with lichens like their equivalents in Devon and Cumbria; perhaps they've only been lost since the Industrial Revolution.

Other threats are only just arriving in Dovedale, although they're more evident to us among the trees. Leaving the truck by the river one still morning, my fellow ranger Kate and I were startled to see two red deer hinds pacing across the ford 50 metres downstream. They paused, knee-deep in the middle of the glittering water, to watch us back, stood amidst sunrays like they'd choreographed the scene. They were beautiful, and their slow pace as they continued up the bank and into the wood held us rapt even after they disappeared from view. These two deer, however, were a bad omen: elsewhere in the country, deer are one of the biggest threats to the future of our ancient woods.

Although red and roe deer are native to Britain, in the absence of their predators like wolves and lynx they've reached artificially high numbers that have tipped their evolutionary role in the landscape out of kilter. Rocketing numbers of introduced deer – fallow and muntjac, in particular, as well as sika and Chinese water deer – cause an even bigger problem. Deer, like so many other species, are creatures of wood pasture. They shelter among the trees even if they venture into fields to graze, but they're just as happy eating flowers and the understorey layer of the woods, not to mention the leaves of any low branches they can reach, and I've seen them balance on their hind feet to munch the nutritious foliage. There are lots of woods, now, where you can see the obvious 'browse line' that we normally associate with the trees in an ornamental deer park; it's possible to see clearly across the whole understorey beneath a ruler-straight base to the tree canopy at the exact height the deer can stretch to. The shrubs beneath are chomped to death and the ground flora is nibbled to nothing.

Parts of Dodgson Wood were a bit like this when I first knew it, thanks to a big herd of red deer that roam the fell above and lay up among the trees, and I was slightly alarmed to see lawns of grass beneath a high canopy of oak in some places and the thinnest young hazel whips nibbled back at the base of the stools. The browsing of the understorey by deer is also implicated in the decline of woodland birds that nest in its thickets, but perhaps most sinister is the destruction of saplings, which is noticeable only by their absence. Like the overgrazing by sheep of most upland woods until the 1980s, too many deer can simply eat the future of a woodland, condemning it to a slow death as the current mature trees decline and fail without any replacements growing beneath them.

Over the past ten years, though, the National Trust's local woodland team has made a concerted effort to coordinate deer control across the Lakes. A dedicated member of staff works with neighbouring landowners and volunteer stalkers to understand the numbers of deer using different areas, then specifies the right culling regime to bring their numbers down to a level that will allow the woods to regenerate – while maintaining the population of animals that are, after all, a natural part of the woodland ecosystem. (This aspiration remains complicated, I should add, by dint of the fact that sheep still get into many upland woods, whether we like it or not, and it can be difficult to tease apart their impact from that of deer. Checking and fixing boundary fences is a tedious and continual task that all too easily falls to the bottom of the to-do list.) The venison produced by the team of stalkers goes to local game dealers and restaurants, although the reliance on people eating game occasionally in restaurants rather than at home was exposed by a massive drop in demand during Covid.

On my most recent visit to Dodgson Wood, I saw bilberry resurgent on some of the little knolls that had previously become grassy, fewer signs of browsing on the hazel and a generally thicker, rougher understorey of bracken and bramble in the open patches, to the point where it – gratifyingly – slowed my wandering progress. Healthy woods are not always easy to get around.

To conservationists, an approach like that of the Trust in the Lakes makes perfect sense, and there are other examples around Britain where deer management works well thanks to a coordinated approach at a landscape scale. Oliver Rackham's mischievous assertion that we should 'eat Bambi' to solve the deer problem doesn't, however, always go down too well with the wider public.[2] The Ashridge Estate in Hertfordshire is home to the National Trust's biggest single woodland holding, more than 1,000 hectares of ancient beech and oak woodland, wood pasture and plantation, but the landscape is also home to a herd of hundreds of fallow deer that have been filmed streaming across adjacent fields. Plans to cull the herd to a sustainable size were met with fierce opposition by local people, who organised demonstrations that were also attended by masked hunt saboteurs, to the alarm of the well-intentioned rangers. With most of British society now disconnected from woodland management, it's perhaps understandable that controlling the destructive overpopulation of wild deer has become conflated with fox hunting and grouse shooting, but it's a very different proposition: in the human-induced absence of predators, we need to act in that role to ensure the health of our woodland ecosystems.

There are alternatives to culling in those places where public pressure makes it too controversial, or where the nature of the woodland, the terrain or the size of the deer population makes shooting unsafe or impractical. Despite us having seen those two hinds, Dovedale falls into this camp at the moment. When I started in the White Peak, most people I spoke to told me there were no deer in the landscape, probably because of the very paucity of woodland and the scale of the open farmland they'd need to cross between cover. Having seen them in the river, though, we also found the signs of an increasing herd of reds using another wood only a mile from the southern tip of Dovedale, where the understorey was heavily browsed and the soil in a corner of the wood trampled where the deer gather during the day. It seems inevitable that their progress into the dale and the landscape beyond will continue and, at some point, there will probably be a need for the type of coordinated management we've seen work in the Lakes. For now their nascent presence was just enough to make us think about protecting our newly planted saplings by using plastic tree shelters, and stringing black netting around a couple of the felled areas where we'd cut scrub and wanted the stumps to spring back. This would all be unnecessary if those two reds were just passing through, but, having invested so much time and effort in the project, we wanted to take a belt-and-braces approach to its success.

Lots of woodlanders do something similar where culling is too difficult, but these alternative measures are imperfect. Apart from the additional effort, expense and resource use involved – and plastic tree tubes are particularly problematic – they also don't solve the issue of deer pressure at the scale of the landscape or even a wood, as fenced areas increase the impact of a big herd on the unprotected area outside. In the long run, we need some sort of more effective local and national strategy for the lethal control of deer, as well as meaningful political investment, whether it results in humans or something else – perhaps reintroduced top predators like lynx – eating more Bambi.

When I mentioned the progress of deer into Dovedale to Joe, he was sympathetic but counselled me to count my blessings;

he and his Natural England colleagues were dealing with an even more challenging issue in Lathkill Dale, where, although it's a National Nature Reserve in the government's care, one of the private landowners still had the right to rear pheasants in pens. Pheasants are native to the Himalayas but have been bred in Britain since the nineteenth century for recreational hunting. They're typically reared from chicks – called poults – in pens inside woods before being released to be flushed out over the waiting guns. The growth in pheasant shooting as a rural pursuit during the twentieth century probably helped save some ancient woods during the Locust Years, giving them a value in the face of competition from agriculture and forestry.

Now, though, many woodlanders have grave concerns about the impact of pheasant pens on our ancient woods: the unhealthy enrichment of the soil is often clear from the patches of nettle and rank grasses around the pens, while inside them the woodland floor is pecked and scratched bare. The birds are also treated with medication that further affects insect populations. Joe and his colleagues investigated the impacts through detailed survey work and were able to categorically demonstrate the highly damaging levels of nitrogen in the woodland soil and severe declines in woodland flora and invertebrates that radiated out from their nadir at the pens.

In recent years there has been a focus in conservation simply on the quantity of biomass that's artificially introduced into Britain's ecosystem through pheasant rearing: around 50 million pheasants are released each year, which, together with red-legged partridges, may constitute a quarter of the entire annual avian biomass in Britain. The British Association for Shooting and Conservation agrees with Natural England that pheasant pens have 'direct and indirect effects on the flora and fauna of the habitats in which they're released'.[3] In theory, the management of a wood for shooting could provide wider benefits for biodiversity – shoots require dense shrubs for cover, open space in the form of rides and glades, and some of the other features we'd like to see more of – but for this pursuit to genuinely contribute to the conservation of ancient woodland the industry would need a very radical shift to a much lower-impact model that doesn't involve raising millions of domestic birds among the trees. For my part, I think that ancient woods are so important for nature and people that we should simply stop using them for recreational activities like this, where the severe detrimental impact is way out of proportion to the number of participants.

The Victorian introduction of pheasants also led to structural changes in many ancient woods. Shrub species like rhododendron and cherry laurel were planted to provide quick-growing ground cover for game birds and for the ornamental value in their explosion of flowers in early summer. These evergreen, non-native species suffocate the woods as they spread quickly both by seed and from layering branches to form an interlocking, dark mass of growth, while the roots release 'allelopathic' compounds that inhibit the growth of other plants.

It's thought that around 100,000 hectares of Britain's woodland might be colonised by rhododendron alone (as well as other areas outside the woods, like on moorland) and, while the biggest areas are in the uplands of Wales and western Scotland, the clearance of both rhododendron and laurel is a priority for conservationists managing woods across Britain.[4] Fortunately, 'rhody-bashing' is a good job for volunteers, requiring only hand tools and often involving a big bonfire, so many of the landowning conservation organisations have made good progress in clearing this threat from their ancient woods. By the time I arrived in the Lakes, for instance, after twenty years of concerted effort by previous rangers few of the Trust's woods there were still home to big thickets of rhododendron and it was a simple task for volunteers to pull the occasional sprigs of persistent regeneration.

That's not to say its removal is easy. I've helped clear rhody from a number of woods, most memorably Jamie Simpson's wood in East Sussex that historically fuelled the medieval ironworks, and it can be demoralising to reflect on your slow progress in hand-cutting each stem and dragging it to the fire when you're faced with a sea of the stuff. Jamie's initial clearance of the whole 10 hectares of Forge Wood, with only one or two willing helpers, was an incredible undertaking but also shows the challenge for private woodland owners: it's difficult enough for charities to tackle species like this with dozens of volunteers, but another thing again for an individual woodland owner to think about doing it.

Rhododendron and laurel aren't the only invasive non-native plant species that can cause issues in our ancient woods. Japanese knotweed, notorious for its impact in urban gardens, can be even more difficult and expensive to eradicate in woodland, while the American skunk cabbage can infest wet woods, dwarfing native flowers among its otherworldly, waist-high plants that look more like giant lettuces. In Dovedale, we're lucky that the only invasive species causing a big impact in the wood is the Himalayan balsam that colonises the river banks after its seed is washed downstream. Like the others, its main effect is to shade out less vigorous native plants as it forms stands of pink flowers in damp places; trusty volunteers, again, patrol the River Dove in early summer to pull the fleshy stems, working closely with the Peak District National Park and other volunteer groups to ensure it won't simply come back from uncontrolled plants producing seed upstream.

Perhaps the most concerning thing about these threats to our oldest woods is that none of them work in isolation; they all compound and complicate one another. On a bad day, in the wrong wood, it can seem like a relentless assault from tree disease, pollution, invasive plants and deer – not to mention the abandonment of traditional management. This can mean, depressingly, that the trees are dying, the special ground flora is suffering, and the whole place is dark and empty of life beyond nettles and rhododendron. Small, isolated woods are even more vulnerable to the combination of threats, with a higher proportion of edge exposed to the impact of surrounding development and agriculture, invasive species more easily dominating, less open space for regeneration, and, even if the invasive species are cleared and the deer controlled, often no easy way for woodland wildlife to return.

Despite the threat of ash dieback, Dovedale is a big, relatively varied wood – certainly more varied than its reputation led us to believe, even if it is also less wild than lots of people think. Apart from the balsam on the river there are no invasive species, and, as we saw, deer only just beginning to explore the valley. As we began the seriously hard graft of cutting ash and planting trees in the White Peak woods, we assumed that we were getting ahead of ash dieback and that some of our new trees might even start to mature as the ash began to slowly die. We hoped, maybe even expected, that the disease would take fifteen or twenty years to have a real impact, as Joe had observed in some parts of Europe, and that our efforts would make the wood more resilient before then.

But we began to see sobering indications of the way the disease would affect the valley much more quickly than anticipated. Checking on our first cohort of freshly planted trees one July day, Kate and I scrambled into the sun on a little limestone knoll where bird's-foot trefoil and black medick grew knee deep. Raised into the treetops, we were confronted at eye level with a perfect example of dieback in a mature ash. Its crown was dense enough at a glance but dead twigs protruded by a metre in every direction, a sure sign of the literal dying back of the tree. Finding such pronounced decline in a fully grown tree felt very different from hunting for signs in saplings, and it hit me hard. The disease was quite clearly well established, and we began to wonder how many other mature trees we'd walked under, oblivious, because we simply couldn't read the crown as well as we could this one.

The disease gathered pace from that point and each summer we'd see its impact more obviously, in shrinking crowns and, increasingly, in whole stands of young ash thinning dramatically so their dominant colour remained grey rather than green, especially where they were already clinging to life on the screes. Once, out with Joe, we even saw the tiny, perfect mushrooms that form the fruiting body of ash dieback on the blackened rachis – the leaf stalk – of dead and fallen leaves. It was a gloomy time for the whole team as we processed the rapid loss. However much we clung to the fact that there were other trees in the White Peak woods, the field maple and wych elm, the remnant lime and the scrub species, it was heartbreaking to watch the wood, as a holistic entity of ash, decline and die in real time.

Occasionally, we'd still meet colleagues who thought we were overplaying things, and we'd take them out and show them what were now the unmistakeable effects of the disease, typically watching without satisfaction as their faces fell. Sometimes they'd continue to rationalise the impact away, proposing reassuring theories: perhaps this isn't ash dieback – these few trees are just dying anyway – or perhaps they'll recover. We'd joke, grimly, that they were in the 'bargaining' phase of grief as they were jolted out of their initial denial.

Our efforts became even more focussed, perhaps even a little desperate – on my part anyway – as if we could stop the fungal tide through force of effort. Dieback became all-consuming as, on top of our conservation project to diversify the woods, we also began to fell and reduce dead and dying ash that were starting to pose a risk to people alongside roads and paths. I realised, not for the first time, that I needed to take care I didn't burn out, and, even aside from the new challenge it offered, the timing was probably good for me to move into my current advisory role that gave me a slight break from being confronted, day in and day out, by dying trees.

*

Unfortunately, worse was to come. In the spring of 2020, the world already felt like a pretty apocalyptic place as we went into Covid lockdown, worrying about – and losing – friends and loved ones. Almost unobserved, at first, except by the skeleton crew of rangers who were still at work during furlough, the ash in the White Peak woods seemed to give up the ghost, whole stands barely coming into leaf while the crowns of legions of others emerged smaller even than their diminished previous state. I stood on Lover's Leap on my first trip to the Peak as lockdown eased and felt a disconcerting mix of emotions as I gazed across at the encompassing vista of Dovedale Wood on the opposite slope; it was exhilarating to be back out in the dramatic landscape on a beautifully clear and mild spring day, but gobsmacking to see the ash take such a sudden turn for the worse.

Over the following months a consensus developed in explanation of this nosedive. The warm spring weather, which had seemed a blessing during lockdown for those of us fortunate enough to have gardens, was actually part of a long period of drought that became acute when the trees should have been starting to draw up water as they kick-started their dormant metabolisms. Simultaneously, there was an unusually late frost that killed those young leaves that were struggling to emerge, knocking back the ash trees yet further.

Having previously been furloughed, I was called back to work to coordinate the response to this sudden downturn across the National Trust's estate. Severe symptoms of the disease were now widespread, I learned, from Cornwall to Northumberland and Suffolk to the Llŷn Peninsula, and many ranger teams had suddenly realised they needed to take action at unprecedented scale simply to mitigate the risk to people. It also became clear that the White Peak's fellow calcareous landscapes of chalk and limestone were most badly affected; not only are they home to more ash trees, but those trees were in the worst condition, probably due to the thin soils that mean they were already struggling for water and nutrients, and so were stressed even before the arrival of the disease. I spent long hours talking with rangers in the Chilterns, Mendips, Cotswolds and South Downs as they grappled with difficult and upsetting safety projects.

As if the situation needed any more doom and gloom, the role of drought and unseasonal frost in the susceptibility of ash to dieback made the theoretical impacts of climate change on our woods uncomfortably real. More frequent and prolonged drought is forecast to be one of the main effects of a changing climate on Britain and its woods, and it's clear now that a long period of climatic stability has ended and that we're already experiencing the impacts of change.

Although drought exacerbated the symptoms of ash dieback in 2020, in the years since we've seen further droughts affect other trees without such a serious complicating factor. In the summer of 2023, the stunted and withered leaves of water-stressed trees were obvious most places I went, and, if the signs of drought are so obvious in the leaves, the damage to the roots and vascular systems will be worse, unseen to us. In the unusually dry summer of 1976 patches of the cambium of many mature beech trees died due to desiccation, allowing decay fungi to colonise them over the subsequent decades, a delayed impact that has seen the development of characteristic cavities in a whole generation of the species. We can only speculate on what might be the long-term impact of not just one year of drought but a whole succession, and we're likely to see changes to woods on thin soils in the south and east of England within my lifetime. (In the east of England, famously dry enough to qualify as a desert some summers, it's already incredibly challenging to establish new saplings due to the almost continual drought conditions.)

Drought is only one symptom of our weather becoming more extreme; we're also seeing more frequent storms strong enough to blow trees over, and woodland soils becoming waterlogged more often from periods of exceptionally heavy rain. British trees are poorly adapted to these polarised conditions as the earth in which they root lurches from being baked dry to a saturated swamp; both can cause root death and a deterioration in the sensitive ecology of the soil. Some tree diseases use waterlogged soil to migrate to the root systems of new host trees, and there's been an increase in ink disease in sweet chestnut in the south-east. Milder winters are likely to lead to an increase even in native invertebrates that can defoliate trees and whose populations normally stay in balance by being wiped out by frosts.

There might be some positives. In unmanaged ancient woods, increased wind 'damage' might actually improve their structural diversity by naturally creating those all-important gaps of light and warmth where wildlife can thrive and new trees regenerate; wind-storms would, after all, have been one of the main drivers of structural change in the wildwood. We saw this happen in many Lake District oakwoods, including Dodgson Wood, during Storm Arwen in November 2021, when big patches were reduced to chaotic, thrilling tangles. When I visited the following spring they were filled with the songs of birds scouting for nest sites in the thickets of fallen crowns. The pits torn out of the earth by the lifting roots had filled with water and I found frogspawn and pond skaters, the water bringing another dimension too often missing from modern woods.

On another positive note, where small-leaved lime was unable to set seed for most of the last few thousand years we now find thickets of saplings as we experience more often the warm weather necessary when the trees are in flower and while the seed is developing. There is one – currently lonely – youngster at the other end of Spring Park from the Anglo-Saxon boundary bank, while I'm aware of carpets of seedlings in other places. Lime is likely to be one of the winners as our climate warms, and could regain its lost place in our treescape over the coming decades and centuries.

These potential good news stories do, however, form a pretty thin silver lining to the black cloud of climate change even within native British woods, and especially in comparison to the bleak impacts already being felt by millions of species and people across the planet. Apart from its direct effects, climate change will act as a 'threat multiplier' to tree diseases, deer pressure, neglect and invasive species, even as they compound one another. Acute oak decline, for instance, is a tree disease that almost certainly has something to do with the relationship between a particular bacteria and a type of beetle, but researchers feel confident that oak trees are more susceptible to infection when they're stressed by the interactions of a number of factors affecting the trees' rooting environment. More oaks display symptoms when we see changes in the soil caused by climate factors like alternate drought stress and waterlogging, as well as air pollution and nitrogen enrichment (with its impacts on mycorrhizal fungi); and certainly, in many places, all exacerbated by livestock gathering under the trees, further disturbing the soil chemistry with their dung while their trampling smashes its delicate structure.

Beyond the impacts on the trees themselves, woodland ecosystems are already suffering the effects of warming as the timings of natural events change, with many signs of spring now recorded weeks earlier than a few generations ago. Most British fauna has co-evolved with other species and they rely on one another for key moments in their life cycles, so these changes can disrupt essential ecological relationships.

The eggs of birds like great tits, blue tits and wood warblers, for instance, hatch at the same time as the emergence of tree-dwelling caterpillars, which the parent birds feed their chicks. The caterpillars, in turn, rely for their diet on oak leaves emerging. All three events – egg-laying, the emergence of caterpillars and bud-burst – are triggered in part by rising temperatures in spring, and, as the warmth has arrived increasingly early in recent years, the three events have followed its arrival and they, too, happen weeks earlier – but to varying degrees depending on the species' ability to respond to such rapid change. Oak leaves now often emerge before eggs hatch, so the chicks miss the peak period of caterpillar abundance, with a measurable negative impact on both the number of chicks that survive and the health of those that do. The birds have also been affected to different degrees, with some better able to alter when they lay their eggs to align with the availability of food.[5] And the very inconsistency of the effects of climate change can make adaptation harder. In 2024, a cold spring meant that caterpillars emerged *later* than birds like wood warblers returned from their winter homes, having appeared too early for them in the few preceding years.[6]

Birds like great tits are 'generalist' feeders who eat a wide range of things, so if they miss the caterpillars they can turn to other invertebrates to feed their young; it's sobering to consider that we're already observing the impact of climate change on these relatively resilient species. Birds that will use artificial nest boxes are also an easier group to monitor and many of the relationships between other groups haven't yet been studied in terms of changes in phenology, but it's thought that the impact of mismatches in timing might be even worse for species that rely on one particular other organism, like the butterflies that only lay their eggs on a specific plant or other invertebrates that feed from a particular flower. On average, British plants now flower a month earlier than they did before 1987, so, if species that rely on particular plants aren't able to adapt so rapidly, the impact will be significant.[7]

In Dovedale, we commissioned an in-depth bird survey before we started working in the wood. The results of that single summer of recording don't tell us much on their own but, if we can repeat it periodically in future, it might help us understand the changes of the bird populations in the valley as the woods change due to ash dieback, as well as illuminating the impact of climate change. We suspect that, in the short term at least, birds might actually benefit as the ash dies, with more light reaching the ground, more decaying wood and, in places, a resurgent shrub layer. On my most recent trip to the wood, when I'd gazed from Lover's Leap on the thinning canopy opposite, the changes felt dramatic as I got down among the trees, with mountain currant and other scrub species responding to the influx of light beneath crumbling ash with rapid new growth. On my two-hour walk I saw four great spotted woodpeckers dashing between the decomposing trees that had King Alfred's cakes and many-zoned polypore bursting from their bark, more in one trip than I'd seen over all my previous visits.

The wych elms that fill the wood in April with the blond glow of their papery bracts also help me consider ash dieback from a longer-term perspective. Until the 1970s, elm would have formed part of the canopy here alongside ash, before the arrival of Dutch elm disease condemned that generation of mature trees. Wych elm is now caught in a constant cycle of youth, seeding at a young age before dying when it reaches the height that the elm-bark beetle – which carries the spores of this other suffocating fungus – flies at. But the wych elms are still here, an important component of the wood. It will only take something to happen to either the fungus or the beetle to interrupt the constant process of re-infection and allow the thickets of young elm to mature again. And this has happened before: unlike ash dieback with its exotic origins, Dutch elm disease almost certainly caused previous crashes in elm that are evident in the pollen record, and, each time, the species has recovered.

Ash, too, will recolonise our landscapes from the small proportion of trees that are resistant to dieback, especially if we can protect our woods from those other threats and give their ecosystems the best possible chance to adapt; it just won't be in my lifetime. It is depressing, in that we've lost, or are losing, the majority of mature trees of two of our handful of native canopy species within fifty years of one another, but also a reminder that populations of trees can and do rise and fall, and that, irrespective of whether we manage them or not, our woods are in a constant process of change beyond human timescales. It brings me peace, in a way, to remember that Gary Snyder, my favourite ecologist and poet, wrote that, 'The woods turn, turn and turn again,' while many woodlanders now refer to 'tree time' as a useful longer-term perspective that helps us think at the scale of centuries.[8] All our woods are different now than at any point in their history, and they'll be something different again – perhaps so different as to be almost unimaginable – at every point in future.

In some respects, the perspective of tree time has helped me come to a sort of emotional acceptance of ash dieback. There are other signs of hope. After that awful crash in 2020, the decline of ash in lots of places has slowed, particularly on richer soils away from the chalk and limestone. As predicted from experience on the Continent, lots of open-grown ash trees – even in places like the White Peak – don't seem to be showing such severe symptoms as woodland ash, and in some cases are even showing signs of what might be recovery. There are a couple of big, wind-dancing field trees on my old commute to Dovedale, their limbs splayed and characterful, that I was dreading seeing start to succumb, but they still looked healthy last time I checked – as I do, carefully, every time I pass.

I'm also conscious of my good fortune in having been part of a community of passionate people responding to this upsetting change; there is strength and solidarity in that shared experience and in acting – working, hard – for a better future. Joan Baez said that action is the antidote to despair and, while I know it's not true for everyone, it is for me – although it's still difficult to walk through the valley and see the impact of the fungus.

And I believe that we took the right course of action in Dovedale, for the woods as well as our own wellbeing. I feel increasingly confident that woods like those in the White Peak could change for the worse without our help; sycamore and beech, neither native to the dales, are most readily poised to move in with their dense canopies casting heavy shade, to the detriment of the special ground flora. Ultimately, though, we don't know for sure how severe the impact of ash dieback on the function of British woodland ecosystems more widely will be; but we do know that the forty-five species of fungi, lichen and invertebrate that are 'obligates' of ash and rely on it exclusively will suffer, while another sixty-two species have a high association with the tree.[9] (Shaggy bracket and King Alfred's cakes fungi are the most recognisable ash obligates, and the rest of the list includes the wonderfully named squirrel-tail moss, Berkeley's earthstar fungus and Atlantic pouncewort, a type of liverwort.)

Above all, I feel sad for ash itself as a species, which I think is an overlooked aspect of the situation. In *Epitaph for the Ash*, Lisa Samson reflects on the trees' loss from a wood she knows well and reflects that, 'The demise of the ash trees . . . is unthinkable, not just for the loss of habitat and shelter . . . but because it is a place of magic and beauty'.[10] Ash may provide vital homes for wildlife and store carbon and deliver all those other 'ecosystem services', but they are also simply extraordinary, unique organisms that have an inherent value in their own right. I still hear some people dismissing dieback because, on the positive side, it will open the woods up, while others feel reassured that different trees will fill the gaps in our landscapes. There is something interesting going on around the value conservationists place on trees as species in comparison to other groups. If we were expecting to lose 80 or 90 per cent of our blackbirds or bluebells over the next couple of decades – or any other iconic and much-loved British species – would we rationalise it away so easily?

The human-induced near-monoculture of ash in the White Peak woods has left them vulnerable to the whole range of threats they face, from tree disease and climate change to the arrival of deer and ongoing pollution. To ensure that their ecosystems are resilient and can continue to function in the face of inevitable change, most of our ancient woods need to be more diverse, with a wider range of their native tree species present and the dynamism – whether natural or human-made – to support the species that rely on them.

There is a wood in the White Peak that is still wilder than even the remnant scraps of prehistory around Dovedale's spires of stone; it is perhaps the wildest wood I've ever been to in Britain, with a mood not dissimilar to the redwood grove that Lodgepole took me to in California. I can't say where it is because there's no official public access, but its location may be obvious with a bit of deduction, and rock climbers are tolerated, so woodland nerds will be too. It sits, like all the area's ancient woodland, on a steep slope, but above a river much deeper and less easy to cross than the Dove. At the top of the wood the whole hill rises up into a great wave of vertical limestone, seeming, when I was stood at its base on a summer's day, more like the Dordogne than Derbyshire.

The lack of a historic local market for wood along with the barriers created by the crag and the river conspired to keep people – and livestock – almost entirely out of the wood throughout the course of history, leaving it as an almost certain remnant of the wildwood and maybe one of the most natural woods in the country. As an example simply of what all the native woods across the limestone dales might have looked like, it is both inspirational and sobering; small- and large-leaved lime dominate the canopy in multi-layered, meandering tangles, while yew and field maple share space with ash, which is present but far from overwhelming. The shifting screes – not to mention lumps of stone that tumble occasionally from the cliff above – create at least some dynamism, smashing trees and making space for scrub and regeneration. It was filled, on my visit with Joe, not just with wildflowers but with the truncated chirrup of redstarts and scratchy blackcap song as we wandered through a unique British landscape.

Joe had taken me there, not just to see a glimpse of the past, but to visualise a future for the rest of the White Peak's woods. With our limited resources and the challenges of the terrain, in the first five years of the project we only managed to fell ash around the other trees species across 5 per cent of Dovedale

Wood, and plant new trees in slightly less. We hope, though, that the resurgent thickets of field maple and oak and alder, and our freshly planted – and carefully recorded – groves of lime, will form new hubs of life and vitality that will spread as the surrounding ash declines and dies. They will rejuvenate this lost world and, in making it a truer representation of the 'virgin forest' of its reputation, also make it more resilient to the range of threats facing our ancient woodland.

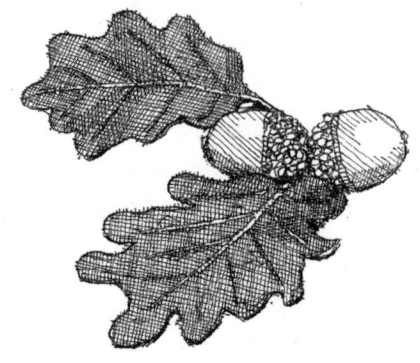

CHAPTER NINE

Inheritance

Generations of oak

A BLUE TIT zips in beneath a spray of oak leaves and works its way along a branch an arm's length from me, hanging by fish-hook feet as it grazes for caterpillars and tiny spiders; life in the canopy doesn't stop for a climber in the way a wood can go quiet when you walk in, but it's a rare treat to get up here with the tree in leaf and be in among its busy inhabitants. We carry out safety work and conservation pruning, like the work to Ashtead's ancient pollards, in winter when the trees are dormant along with much of the life they support, and with their crowns bare the sky feels enormous and exposing. Now, though, the foliage defines intimate, room-like spaces among the branches, shady and enclosed. Slivers of sun pass between the leaves and pick out filaments suspended vertically with tiny, part-translucent green larvae spinning slowly from the bottom. I've never knowingly seen a winter moth, which each little caterpillar will become, but the grubs are ever-present beneath the oaks in early summer and I'll shake dozens from my shirt and hair when I get back to the ground.

Spring Park hosts a few dozen quietly spectacular oaks, the 'standards' that were historically encouraged to grow above the hazel coppice. While each coupe of hazel was cut every eight or ten years, these few select oaks were encouraged to grow on as timber trees over multiple cycles of coppicing. There were just one or two in each coupe so they didn't shade out the hazel regrowth, and while some might have been felled after seventy years for local use – framing for houses or barns – others could have been left to grow older to provide the beams for grander architectural projects.

Tending the oaks was a remarkable multi-generational task, with each woodlander making nuanced decisions about which trees to prune or fell and where to allow the development of new standards, and they had to provide for their own needs while ensuring a steady supply of timber for their descendants. Oak holds the most obvious immediate potential for a resurgence of this careful, long-term production that could help bring our ancient woods back into good management, although new technologies that incorporate lower-quality wood into effective building materials offer opportunities for the use of other species. A deep knowledge of oak, in its many forms, is inescapable for any British woodlander, and each one I climb or work with adds another layer of depth to my relationship with the British landscape.

The cohort of big oaks left now define Spring Park despite their relative scarcity; there are perhaps only thirty or forty trees across the wood. When I first arrived here their towering stems, each subtly but unmistakeably unique, provided key markers for navigation on the grid pattern of tracks, determining the particular atmosphere and aesthetic of their location. I quickly came to know them individually and they became a familiar, constant presence as we coppiced the hazel and chestnut surrounding them. These trees must have been almost at their peak as a crop when the wood came into the City's care in the 1920s, and some of them are as big as woodland oaks in the south-east are ever likely to get. In that last century they've gone past the point of being the prime timber trees they undoubtedly were and into the realm of the 'veteran', developing characteristics that encourage their use by an increasing range of other species: broken branches, pockets of crumbly decay, woodpecker holes and skeletal deadwood, all busy with life. Spring Park's oaks don't have the huge crowns of Ashtead's ancient pollards – and neither are they as old – but they're special in a different way, the result of coppice management rather than wood pasture.

The first part of any climb up these giants begins with an arm-aching haul, scaling the rope past 10 metres of straight stem without a branch to swing a boot over. As a young tree, the surrounding hazel would have drawn the oak up to the light, suppressing its branches and creating the clean trunks that were of the highest value when sawn into straight, knot-free timber. Once the oaks emerge from the shorter hazel canopy, the crowns typically 'break' and chunky limbs arc away from the main trunk. A few of the trees here, though, continue unnaturally straight and tall given the fact they'd have only been surrounded by hazel during their formative years. They might have been pruned up further during their youth to make even better timber, the woodcutters free-climbing them or using long ladders to saw off rogue side branches.

Oak timber from trees like these looms large in the British imagination. We think of the 'hearts of oak' of Nelson's navy and iconic timber-framed buildings, from black and white Tudor pubs to cruck barns and the soaring vaults of cathedrals. Oak is one of the most common species of canopy tree in the UK and almost certainly the best known; it would be easy to think that its prevalence in many places is entirely natural. While it would certainly have formed an important component of the wildwood, the species was actually favoured and 'promoted' by our ancestors purely because it was so useful. In the uplands – like the temperate rainforest of Dodgson Wood – it was often managed as simple coppice for its bark and to make charcoal, while in the lowlands it was much more commonly grown as a standard tree for its timber above other species of coppice, as we find here.

Oak was the preferred timber for construction in Britain from the point that people began to build permanent structures, which we've been doing since before we learned to farm, until the mass importation of cheaper and more uniform softwood began in the nineteenth century. It's strong, behaves predictably as it dries, and contains a high level of tannic compounds, which resist colonisation by decay fungi when the tree is alive, and also mean that the wood is incredibly durable once in use, lasting hundreds of years. Even where it's exposed to the elements or in contact with damp earth it will last decades where the timber of most other British trees would rot in a few years. Preserved oak has been recovered from the remains of Bronze Age roundhouses, where it was used for the posts that supported the roof, while the walls were of wattle and daub made from hazel and willow withies.

The Romans introduced a level of sophistication to their oak buildings that included sawing and squaring the timber and the use of simple mortice joints; a protruding tenon was cut on the end of one beam that fitted into the chiselled mortice on the other, where it would be secured using iron nails. In the Early Middle Ages that followed the end of the Roman Empire, British construction reverted to its prehistoric technology with the use of round timber and simpler joints, until a cultural and technological revival began around the twelfth century that saw a resurgence of highly sophisticated timber framing.[1] Squared timbers were connected with increasingly strong joints, the frames worked 'green' before the wood dried so that the structure would become even tighter as it warped slightly over time. The use of oak pegs – sometimes called 'trenails' or tree nails – to secure the joints was an improvement over the Roman fixings as the pegs were similarly locked in by the movement of the wood, while thin iron nails would have been quickly corroded by the tannins in the oak. You can see this effect today where, when ironwork has been fitted to a new oak gate or door, the chemical reaction between them forms a black stain that lasts for years before the oak fades to its permanent silver hue.

The sheer quantity of oak that was needed for building from this point on is evident if you visit a medieval cruck barn like the monastic barn at Great Coxwell in Oxfordshire or the tearoom in the barn at Cotehele in Cornwall, where you can admire the timber framing with a scone. Unlike many less utilitarian buildings, the intricate but substantial wooden structure of the barns is exposed so you can see the thickness, length and sheer number of timbers necessary to enclose these cavernous spaces, not to mention the functional elegance of the joints. As well as their rustic beauty they have a numinous atmosphere despite their core purpose as agricultural stores, and it's no wonder that they were also used for ceremonies and celebrations.

Oliver Rackham, meanwhile, counted beams from around two hundred and fifty oaks in a middle-class 'hall house' at Flatford in Suffolk, varying in size depending on their use. He calculated that a house like this – which would have been fairly common – might have used the entire annual harvest of standards from a coppicewood of 61 hectares, three times the size of Spring Park.[2]

But special as they are, timber barns and houses are dwarfed by some of our most significant buildings of the same era. The restoration of Notre-Dame after its catastrophic fire provides an insight into the amount of timber forming the roofs of equivalent British churches and cathedrals. More than 1,000 oaks of 150 or 200 years of age have been cut for the new roof, some of them providing single timbers of more than 25 metres in length (which it would be challenging, at best, to source in England today).[3] Artisan carpenters who've been involved with a unique project to build a medieval castle from scratch, using only contemporary techniques, have been drafted in to inform the replacement of the original roof, which was installed in the thirteenth century and survived until the fire.[4] Closer to home, I climbed into Ely Cathedral's Octagon Tower, which relies on eight oak beams of 20 metres in length along with 400 tons of other oak framing hidden behind its elaborate décor. One of the wonders of the medieval world, its suspended structure is supposedly not fully understood by modern architects, nor the way it was erected. The guide also reminded me that its wide roofs would originally have been tiled with tens of thousands of wooden shingles.

These tall, straight stems of oak, that it takes such effort to climb, were often processed where they fell to avoid the need to haul waste, and in some ancient woods we can still find the remains of medieval saw-pits, short trenches above which the fallen butt would have been laid. The sawyers laboured with a double-handled saw, one on top of the log and the other – the underdog – in the pit below, showered in tannic oak-dust. I like the smell as the damp flakes of oak plume out of the back of a chainsaw; it is pungent and sharp, but not unpleasant to anyone who appreciates builder's tea or an earthy red wine. I sympathise, however, with the underdogs, who wouldn't have had any protective equipment beyond a wide brim to their hat – I was once hospitalised when a rogue chip of sawdust made its way under my visor and into my eye.

It's a measure of the importance of a sharp saw that, where logging took place in camps on the west coast of America, the 'saw doctors' devoted to maintaining the tools were some of the best paid woodland workers – while the other top-paying job was as cook, which should be no surprise to food-obsessed rangers. Today, most woodlanders prefer to sharpen their own chainsaws, believing that nobody else can quite do it to their exacting standard, but it remains crucial to the job and a saw with the edge taken off by accidentally dabbing soil or stone can make hard work of the easiest cuts. Saw-pits have become a legendary part of British woodland history but they're actually a relatively recent invention; before the fifteenth century timber would have been cleaved apart using wedges and squared – or 'scappled' – with an adze, a tool like an axe but with the edge set perpendicular to the handle. Clive Mayhew, the expert on woods and war, pointed out that in the Bayeux Tapestry, painstakingly embroidered in the eleventh century, there are lots of depictions of people working wood but no saws, which, although they existed, weren't yet good enough to efficiently process big stems.

In comparison to some aspects of traditional woodland management, the economics and practicalities of the oak timber industry are pretty well understood thanks to the diligent research of landscape historians and, because of its significance to powerful people, there are comprehensive records from many places. Beyond their lists of products and prices, the estate accounts just begin to bring to life the network that operated over generations as the biggest oaks matured, across lowland Britain and into mainland Europe: landowners and their agents, timber buyers, shipbuilders, architects and engineers, woodcutters, sawyers, carters, horsemen, carpenters and joiners; not to mention the wood colliers, bark-peelers, furniture bodgers and other coppice-workers. Even the movement of all this oak between the woods, towns and shipyards was a convoluted logistical operation by horse and boat that sometimes saw the biggest beams take years to reach their destinations.

While oak was the most important timber tree – George Peterken writes that, in some medieval records, 'standard' is synonymous with oak[5] – other species were also grown above coppice for specific reasons: ash for flooring and cart-building, hornbeam for chopping blocks and the giant cogs in wind- and water-mills, beech for furniture making, and field maple for musical instruments, all among many other uses. Perhaps the best example of the high level of refinement in our ancestors' use of wood is in the construction of carts and wagons, essential from their invention until the development of the combustion engine. Nearly every component was made from a different species that best suited the particular strains it would be subject to. Elm, for instance, is solid enough to form the hub of a wheel and won't split even after having a number of mortices chiselled in to take the spokes, which were more often of oak, which stays strong over their relatively thin, long lengths. The 'felloes' that made the rim of the wheel were of springy ash that could absorb impact and formed a limited form of suspension, while the brake pads were cut from heat-resistant black poplar.[6]

Stomach straining from the effort of swinging and heaving myself up the rope, I finally hook a heel over the first branch I come to, a fat, mossy limb that I roll onto for a rest after a graceless scramble. Above the height of the hazel, the crown breaks and the tree spreads out – and even down – as well as up. Some of the branches are bigger than I'd be able to wrap my arms around and shoot off horizontally, cantilevered improbably for metres. Others zigzag and twist to make the architecture of the crown convoluted and intriguing as the branches pursue the light in all directions. It's a very different climb now, the wild, self-willed structure forming a giant climbing frame. Instead of pulling myself up the rope I let the tree guide me from branch to branch, stepping from bent elbow to dipping knee, traversing beams to find the next point of ascent and quickly winding my way up.

These tangled crowns also played an important role in the old woodland economy; the crooked limbs of oak were sometimes known as 'knee timber' and were used to make strong curved beams without being joined or cut into shape. They were harvested for 'cruck blades', the roof beams that give those barns their name and which were often produced from the two matching sides of the same beam, as well as for the framework of boats, where their natural curve would have again been matched to the exact angle needed. It's thought that many trees were pruned and trained using props or ropes to create the most useful shapes; in the oakwoods of the Basque Country in northern Spain – an extension of the temperate rainforest of Britain's west coast – whole groves of open-grown trees display contorted shapes as a result of their history as *ipinabarro*, which loosely translates as 'guided trees'.[7] Their knee timber was destined for dockyards on the coast and, like ancient woods in Britain, they survived because of their importance to this industry – which, pleasingly, then enabled the development of a human connection between these two matching ecosystems as oak boats crossed the Bay of Biscay.

Boats were built of rough-hewn oak planks fastened to a frame of naturally curved 'futtocks' – the ribs supporting the hull of a ship – from as early as the Bronze Age, but it was during the same technological renaissance that saw the resurgence of timber-framed buildings that big ships began to be constructed using large-diameter knee timber. This began the era of Britain's naval dominance that eventually saw iconic vessels of oak like the *Mary Rose* launched in the sixteenth century and HMS *Victory* commissioned in 1778, ready for its role as Nelson's flagship at the Battle of Trafalgar. Detailed records show that 6,000 trees were needed to build *Victory*, with around 90 per cent of them being oak, while elm was used for the keel and pine was brought from New England for its masts.[8]

Britain's military expansion during this period caused competing demands for oak. The increasing production of cannons in the sixteenth century, for instance, created a greater need for charcoal to fuel ironworks, forcing Henry the Eighth to issue the Act for the Preservation of Woods in 1543, which required woodland owners to retain twelve standard trees per acre for use as timber. Further royal decrees followed over subsequent centuries to try to ensure that enough oak was being grown to supply the navy, with particular concerns persisting that the woods weren't being restocked with new saplings.

The expansion of the Empire and the ever-increasing need for long-distance merchant vessels led to the heyday of oak shipbuilding from the end of the seventeenth century until the middle of the nineteenth. Swathes of common land in places like the Forest of Dean were enclosed and planted with oak during this boom to fulfil the assumed future need for naval timber, and half of all wooden ships ever built in Britain were launched between 1800 and 1860.[9] Advances in technology and the rapid success of ironclad warships, first used in anger in the American Civil War, marked the onset of the use of metal for the frame and hull of most new boats and saw the end of any meaningful demand for British oak for shipbuilding.

Throughout this 600-year period when oak played such a central role in Britain's story – and consequently in that of many other places – the 'woodwright' was at the heart of the economy and that extensive network of trades that kept the woods alive with human life. It was the woodwright who visited coppicewoods to choose and buy standard trees or even individual branches that would suit specific needs, sometimes years in advance of their felling and use, making considered judgements of the knee timber high in the crown in another essential aspect of what came to be known as woodmanship. Having selected the right trunks and crucks, they would organise their careful harvest, with giant boughs sometimes being lowered using pulleys to ensure they weren't damaged during felling, and mats of faggots – the bound bundles of twigs produced from coppicing – laid to cushion the fall of valuable trunks. As I step lightly from bent limb to crooked bough I find it easy to appreciate the architectural value that could be revealed from each bend and turn of the knee timber; harder to think about the reality of breaking this giant structure down with a felling axe, or of linking the angle of each joint with the particular boat keel or roof-arch it was destined for.

One of my most treasured books on the traditional uses of timber and wood is *The Shell Book of Country Crafts* by James

Arnold. Like the Shell guides to English counties (to which John Betjeman, Paul Nash and W. G. Hoskins contributed), it's a detailed but lyrical account of dozens of industries, many of which he managed to catch in 1968, just before they collapsed completely. Arnold notes of the woodwright that wooden goods were 'built by men who went out to the woodlands and bought their timber by direct selection. They knew that years would elapse before this timber would be . . . ready for use. Generations of practice had produced a type of man and engendered a way of thinking in wood. Into that making went something else that is hardly understood now.'[10]

I would never claim to be able to think in wood, but I have seen that 'something else' surviving in people I've been fortunate to meet. The past couple of decades have seen renewed interest in building from green wood, and I had a memorable visit to Black Down, on the border between Surrey and West Sussex, where I met the National Trust lead ranger, David Elliott. He's been working the woods there for nearly thirty years and, after a diffident start, spoke in an incredibly moving way of tending the oak standards over years before occasionally choosing the right one to fell, which is then milled from butt to branch for floor- and weatherboards, beams for timber framing, and countryside uses like gateposts and benches.

It was obvious that he took the responsibility of this decision seriously but that the use of the timber from each oak tied him and his team ever deeper into the story of these woods, the beautiful buildings and products they create binding them to the surrounding landscape – where their structures look completely at home. There's a wood-framed house of chestnut poles, constructed with the help of Ben Law, which long-term volunteers can use while they gain essential experience. Perhaps even more beautiful, though, is an apple barn and workshop made using chestnut and oak framing techniques.

David showed me the giant jig in a corner of a field where the rangers set pairs of interlocking beams at the correct angle to chisel their tenon and mortice, painstakingly constructing the frame in two dimensions before its eventual erection like a giant 3D puzzle. Apart from David's deep relationship with the active management of the woods – which also sees areas of old coppice being brilliantly managed for the benefit of wildlife – what stuck with me was how heavy and difficult to manoeuvre the beams were even for the relatively small apple store. The rangers used a tractor to place them on the jig and then to build the final structure, but before their invention woodwrights depended on horse- and human-power alone, with just the use of pulleys beneath temporary tripods of poles to help erect frames or to lift logs onto carts. It's easy to see why grand buildings like cathedrals and castles took decades or even centuries to complete.

We have long, friendly debates about whether we should fell any of the big oaks at Spring Park, like David has at Black Down. While it's important to knock out some of the taller trees over the coppice to ensure enough light reaches the freshly cut coupes for nature to thrive, it's always easier – both practically and philosophically – to take the younger generation of canopy trees that don't hold such individual interest. But if we always do that, we might not leave a successor generation to grow on to replace the giants when they inevitably collapse. We never reach the answer during my time; these trees are probably a particularly difficult group as they're all of the age where they've developed lots of important habitat features in their broken stubs and pockets of decay, and it's important to retain them when we can.

Choosing which big oak to harvest every five or ten years – or more – isn't the kind of decision it's easy to make without knowing a wood for decades, even when it's populated by less characterful trees. Nor is this kind of generational thinking easy to record in a management plan, which is an unavoidably short-term statement of intent in comparison to the life of an oak. There are very few woods now – away from the biggest family estates – where we can be sure that our successors will continue our traditions of management once we're gone, especially without the consistent, long-term demand for particular types of timber that our forebears experienced for centuries.

After years of caring for Forge Wood, clearing it entirely of an infestation of rhododendron and reinstating the coppice regime, Jamie Simpson decided to harvest a few of the bigger oaks to make more space for regeneration. I was in the habit by this point of regularly helping him out, typically in return for patisserie and rich little quiches we'd stop to buy in Heathfield, where the bakery looks like a typical small-town English shop but has an exceptional French *boulanger* churning out the best pastries I've ever eaten. (Following one catastrophic visit when he hadn't baked any of our favourite glutinous almond croissants, the baker gave us his card so we could ring up the day before to tip him off.) Jamie invited me along for the big day when his mate Andy, who usually did the 'forwarding' work, pulling timber out using a tractor, had offered to do some of the felling and help measure the oaks for sale.

It was amazing to watch him work. Both Jamie and I had seen big trees felled before, and cut plenty ourselves; in fact we'd recently removed a couple of dozen giant, non-native turkey oaks from another wood. But, like a medieval woodwright, Andy knew some particular techniques to soften the fall of the oaks and ensure the main trunks stayed intact to retain their value. He contemplated each tree from every angle as well as the lie of the land around it, at one point asking us to fill a steep little ditch with brash, before making his felling cuts with exquisite care, sighting and re-sighting the gob – the wedge cut from the front that dictates the direction of fall – then tickling the back cut in so that the movement of the crown was barely perceptible until the tree reached a critical angle and crashed down.

Once we'd snedded the crown, trimming the side branches off, Andy showed us the correct way to measure the straight trunks and calculate their volume using the forester's 'blue book', which takes into account the diameter and taper of the stem depending on the species. It was an uncommon and slightly bittersweet experience to count the rings across a metre of damp, golden wood at the foot of the fresh stems, the air ripe with tannins, and to lean on the waist-high log as we gathered round to watch Andy jot figures in his notebook. It felt timeless, and I wondered how often this scene had been repeated through the history of this industrial old wood. In Jamie and Andy I saw modern-day woodlanders at home in their natural habitat, and, while Andy is from a long line of rural Sussex folk, what's heartening is the acceptance with which he and his colleagues have welcomed Jamie, then a young Londoner with a range of colourful experiences, but who's won their respect through undeniably hard graft and his commitment to improving the place for nature.

Jamie's oaks would fetch around £1,000 each from a local sawmill, where they were milled into beams for the maintenance of old buildings – typically things like replacement lintels – and for gateposts, a sharp contrast to the couple of hundred pounds he could expect for a similarly sized tree going for firewood. Even that price, Andy told us, was a scandal. The timber should be worth much more but won't make it into any new building projects despite being of high quality: straight and free of knots.

The British forestry industry faces a range of issues that conspire to keep British oak out of higher-value markets. Not much of it is being grown and harvested to a high standard, for instance, which means that there's little incentive for sawmills to have it graded to meet building regulations – which further reduces the interest in growing it. The result is that 98 per cent of the oak timber used in British building projects comes from France, where it's mainly grown in more closely tended plantations and has an entire industry around it; those 1,000 trees for Notre-Dame represent a tiny proportion of the French annual harvest.[11] There's an exciting opportunity to develop a home-grown supply of quality oak by bringing more of our ancient woods back into management, as well as by developing new – nature-rich – plantations. It requires those various issues to be resolved, though: we need to develop better knowledge and skills among woodlanders so that they can grow quality oak and harvest it at the right time; increase confidence and willingness from the sawmills to process and test it; and, of course, encourage demand from architects and builders to make it all worthwhile.

Growing more oak timber will also require the control of grey squirrels, which are often spoken of in the same breath as deer in terms of the threat they pose to British woods. Greys were introduced to Britain in the Victorian era as a novelty but have spread almost everywhere and driven our native red squirrels to a few, isolated refugia. They also cause catastrophic damage to timber trees by stripping the bark on the leading stem and branches, preventing straight growth and creating stunted, irregular specimens.

Unfortunately, controlling grey squirrels can be even more challenging than managing deer. Even in those places where shooting is physically – and politically – possible, it must be done at a wide scale or more squirrels from the surrounding area just fill the vacuum created. Traps are available but lots are needed to cover bigger woods and checking them quickly becomes onerous, while there are no physical protection measures equivalent to tree tubes and fencing. It's therefore typically only worth trying to control squirrels – normally by shooting *and* trapping – where the high costs of doing so will be worthwhile in relation to the increased income that quality timber will generate, although many in the hardwood industry have their hopes pinned on a contraceptive for the so-called 'tree rats' that's currently being field tested.

Most of the trees felled as part of the management of our ancient woods today end up as firewood, which is a depressingly low-value end for timber that would have once been integral to so many other aspects of life, but is perceived to be the only use for the variable wood of our native broadleaves. Jamie has sold hundreds of tons of overstood coppice for fuel, alongside claiming whatever government grants he can (and even between them his management of Forge Wood, primarily for nature, pays him a meagre hourly rate considering the amount of time he's invested in getting the detail right – like cutting pollards on the boundary bank).

While the production of firewood can just about sustain the active management of woods that have good access for extraction, I suspect that the days of relying on its sale to support conservation work are numbered. Wood fires – and even the more efficient stoves – are one of the biggest sources of particulate pollution both inside houses where they're fitted and more widely in towns and cities. It's clear that we need to drastically reduce their use, particularly in urban areas, a fact brought home to me by the decision of a tree surgeon I know to immediately close down his lucrative sideline selling firewood having read his daughter's school project on the subject. There's already an increase in legislation designed to limit the installation of new stoves and to control the type of fuels that can be burned.

I recognise how difficult this is: like many people, sitting by a fire is one of my favourite simple pleasures. But as we urgently adapt to a low-carbon future, we'll need to figure out how to live with burning much less wood – as part of our essential drive to stop burning anything – while still accessing the rich human meaning and connection that comes from communing with others as we stare into the flames. The cultivation of fire is, after all, our defining, original skill and of deep psychological importance. But for woodlanders I worry that this will become simply a practical challenge of finding new markets for their wood – with a risk that without alternative sources of income some woods might fall back into neglect.

Britain is currently the world's second biggest importer of timber, shipping in more than 20 million tons each year, while we produce around half that (much of which is exported) – so across all types of woodland there's huge potential to increase what we grow ourselves. The majority of the imported wood is softwood framing and ply for building, wood pellets for biomass power stations, and paper or wood for pulping, so our ancient woodland could never get anywhere close to replacing it like for like.[12] There are, though, other avenues we could explore if we can solve some of those knotty challenges around supply and demand, particularly as the construction sector looks to innovative wood products to replace carbon-intensive materials.

Advances in the way wood can be processed and converted to the most useful form for building mean that modern 'engineered timber' – like laminates, where strips of wood are glued together – can now be used as structural components, so smaller and wonkier pieces of wood can be brought into use in bigger end products. The beautiful winter garden near me in Sheffield, a giant greenhouse sheltering tropical trees, is more than twenty years old and its soaring glass roof is held aloft by curved beams of laminated larch, each 20-metre length built from much shorter sections. The technology has moved on since then so that even shorter and less straight timber can be combined into useful beams, and recent years have seen the construction of big wooden buildings like a graceful seven-storey office-block in East London. The Black & White Building has a frame of laminated beech that's stronger than steel, while its construction resulted in less than half the carbon emissions of an equivalent concrete structure and continues to store more than 1,000 tons of carbon.[13]

Elsewhere, knee timber has been employed in trial projects using computer modelling that quickly incorporates a range of irregular natural curves into a holistic design. Specialised heat treatment can be used to make the wood of sycamore and ash suitable for external use, when it could only previously be used indoors, increasing its ability to replace imported softwoods for things like cladding.

There are similarly encouraging innovations happening in technologies that break wood down even more comprehensively; chemical processes are applied to extract its key organic compounds, cellulose and lignin, which can then be repurposed into textiles and other materials. As all these technologies become increasingly mainstream, there's no reason why they shouldn't utilise the wood harvested from neglected coppice during its restoration, or indeed the ongoing produce of in-cycle coppice – with careful thought and if we can link the right parts of the supply chain. It seems unlikely that the timber markets of the future will rely solely on trees that grow straight and tall, as they have done in recent years, presenting an opportunity for our ancient woods to produce essential materials once more.

Biomass boilers and power stations, meanwhile, form a cleaner alternative to domestic firewood as they burn much more efficiently and with less pollution. Many were installed with the intention of using local wood, and sweet chestnut from Spring Park has periodically been chipped for this purpose. Skewed global economics, though, mean it's now cheaper to import wood pellets from South America and Malaysia – often contributing to deforestation – than to burn chip from sustainably managed British woods, particularly once the additional cost of using sensitive extraction techniques rather than giant harvesting machines is factored in. Even within the National Trust, there are some places where smaller-scale biomass boilers have been installed to heat historic houses, offices and tearooms but where simple economic reality means it doesn't make sense for them to be fuelled from local Trust woods, which is frustrating to say the least. But there are some success stories. At Croft Castle in Herefordshire, motivated and creative woodland management means all the fuel for the boiler comes from the estate, and illustrates the potential for more ancient woodland to provide a less polluting form of woodfuel.

As modern woodlanders pursue these opportunities, we mustn't forget the needs of future generations, or indeed the health of the woods. Timber trees should only be harvested in ancient woodland where we're simultaneously restoring the coppice, with the encouragement of a new generation of standards, or where well-planned selective felling ensures that production is part of a genuinely sustainable system that maximises the protection – and resurgence – of woodland wildlife. Soils and archaeology must be carefully protected and heavy machinery only used where it won't cause any damage. I was once asked to visit an ancient wood in private hands where perhaps a quarter of the oak standards had simply been stripped out in one go, and the forwarder had created ruts a metre deep in the previously untouched earth. The owner proudly told me that the tens of thousands of pounds they'd been paid for the trees partially offset the purchase of the wood, and that they would make their money back in full if they could do something similar a couple more times in their life; but with no thought given to the nurture of new saplings, there was no new generation of standard trees developing and the future of the wood looked bleak.

Although the climbing is easy across the rough tangle of boughs in my big oak at Spring Park, I reposition my ropes every couple of moves to protect myself from a fall. The irregular framework of thick limbs makes each of the big oaks I've climbed, both here and at Ashtead, a world of its own. I've come to understand that I have a wordless map of each one drawn in my subconscious without ever being able to verbalise that knowledge when I'm back on the ground.

There's a familiar messy drey – a squirrel's nest – tucked in an old wound where a branch broke out, and a long crack in a giant limb that has partially given way; perhaps it's now a home for bats. A line of woodpecker holes runs up a thigh-sized branch, hidden from passers-by on the ground. Once I knew they were there, I paid more attention when I walked by, and in early spring saw the undulating flash of a great spotted woodpecker homing in. Then the pay-off: woodpecker chicks clamouring for food, sounding like squeaky bike wheels. Today, otherworldly sponges of chicken of the woods fungus billow from the bark between two of the holes, a clue to the decay within that has allowed the birds to excavate their nest. The layers of succulent bright-lemony frills will turn chalky and crumble away before most of the other fungi fruit in autumn.

Climbing forces you to zoom in even further on the detail of the tree, an immersive macro mode. I swing a leg above me and over a big limb, then haul the rest of my body after it until I'm lying face down with my cheek on the rough bark, looking through a forest in miniature of moss and epicormic growth: soft and thin oak twiglets. Just a few centimetres tall, they could be years or decades old, restricted as they are by the shade of the main crown. Sometimes they're a sign of stress, an indication that the foliage is thinning and the dormant buds beneath the bark have been stimulated by the increase in light. Lots of oaks, though – like this one – are just 'furry', and it seems to be a genetic trait that epicormic shoots grow all over the stem and branches, even when the tree is in good health. Miniature groves of sporophytes, the fruiting bodies of mosses, stand up above the green carpet between the twigs, an understorey to their tiny canopy, while, after a few moments' pause, I begin to see woodlice scurrying between plates of bark that hide tiny, spherical spiders' eggs.

Moments like this have helped me learn to better appreciate nature by stopping and looking more closely; it wasn't that I didn't have the patience before, but the extreme physical intimacy that climbing – and felling, and lots of other woodland work – brings with the life of the woods forced me to realise how much more beauty and detail there is to be found when we come nose to nose with it. Proper ecologists know this, of course: the identification of many plants is dependent on the inspection of tiny details using a hand lens, and there are beetle species that can only be separated following an examination of their genitalia under a microscope. I've become a passable naturalist of the habitats and species I've spent the most time with, but of bigger importance to me is the deep, almost workaday knowledge I have of a few key species – like the oak.

Many – perhaps most – people can identify an oak tree. But in the same way I've learned the crowns of the trees I've climbed, I realise that years of being forced so close to oak – and so physically – have brought me to know the species itself in a way I find hard to articulate but which is inherently, and constantly, within me. The groves of twigs and sporophytes among the moss live in my subconscious with the crusty lichens on the smooth outer branches, while I carry the feel of the rough bark on the bigger stems, and the way it changes as I climb, in my palms and fingertips. Thoughtless observation of the changing foliage marks the progress of each summer as bright new leaves, lush and fresh like seaweed in a rockpool, toughen into a dark late-August canopy that takes on a blinding gloss in the morning sun.

There's more than I can bring to mind: the heft of oak brash, the smell of leisurely decay from a wound left by a broken branch, the little buttons of spangle galls beneath the leaves, the twists and turns of the branches, unique to each tree but always so *oak*. It is a world in itself, contained within those three letters: these standard oaks at Spring Park could almost be a different species to Ashtead's pollards. The gnarled and twisted trees at the top of Dodgson Wood, splattered with leafy lichens, are, of course, a different species, the sessile oak of the uplands rather than the English oak, although the differences of shape and character between the two are overwhelmed by the influence of their environments and the long human histories that have driven the diverse forms of this tree. I almost take it for granted now: but when I do stop to reflect I realise that my unthinking understanding of the oaks I walk among most days has become part of who I am, rooting me in this place but also in the oak-dominated landscape of much of England, from the coppicewoods of the south-east to the rainforests of the Lake District.

Nearing the top, I pass the last fork I'd really, truly trust my life to; on an oak, dependable as it is, the two branches are about the size of my wrist. I leave my rope attached there, pulling the slack after me, and climb on, barely thinking. This is the muscle memory of tree climbing, learned at five and never left behind. I'd be nervous without a rope though. The thin outer branches, deep red or silver-brown and not yet furrowed by age, flex, and it takes a few quaking moves to find a perch where I'm poking my head out beyond the last leaves but solid enough that I can relax and strop on with my lanyard, a short supplementary rope that frees my hands as I gaze out. With the slope facing south-east, you wouldn't know you're on the edge of London. The view across the billowing green of the treetops is of fields of crops and horses rolling away towards Biggin Hill and the Kentish Weald.

A movement closer by pulls me back, a little blur of dark silver in the corner of my eye, a flash of colour – purple? It takes a moment to focus, then I know it immediately, despite never having seen one this close: a purple hairstreak butterfly fluttering out a few inches from the tree when the air is still and dashing back into shelter when the slightest breeze bothers the leaves, sometimes not even enough to rock my slender branch. There are more, two or three at a time tumbling out end over end and over each other, in loops and spirals. Suddenly, I realise they're everywhere, the whole crown exhaling and inhaling with dozens of the tiny butterflies, keeping time with the gentle currents of the summer air. The irregular rhythm is hypnotic; the visual effect trippy and reinforced by the soft, sympathetic sway of the branch I'm tied in to. We seem to breathe and move together, the tree, the butterflies, and me.

Later on I look the hairstreaks up. I've seen them from the ground before but only ever a couple at a time silhouetted at the crown's edge, and I had no clue what might be going on out of sight up above. The book tells the same story: 'It is possible to walk beneath a tree bearing several hundred purple hairstreaks without being aware of their presence.'[14] I find I've been even more fortunate, as the butterflies' numbers go through dramatic cycles, from a couple of pairs on a given site one year to thousands the next. And they don't use every tree but congregate on particular specimens; nobody's quite sure because they're so hard to study, but it's perhaps those oaks that aren't quite the tallest but offer a broad crown that's both sheltered and unshaded.

I lose minutes – who knows how many? – before my climbing partner, Sean, shouts from below: 'Alright up there?' I don't think he's too worried; I can just make him out sat on a fallen branch, smoking. I give myself a couple more breaths with the hairstreaks in our private, hidden world, opening my chest and moving with the tree as the silver-purple confetti flutters out around me and retreats again. They're so tiny they'd be lost against the canopy from even a few metres away, and I imagine the lot of us disappearing into insignificance against even this small wood, never mind the wider landscape. As I drop gently between the big limbs I wonder whether there are any other trees at Spring Park home to this secret, magnificent spectacle, and what else happens up here, invisible to us as we pass by below.

In a few places across Britain's ancient woods, walkways and towers allow visitors easier access to the canopy and, at Sheringham Park in Norfolk, I once watched a handful of purple hairstreaks tearing around above the oaks from the 'gazebo', a modern folly built in the 1980s. While there's a part of me that enjoys the privileged, sometimes unique experiences of nature that woodland work brings, I believe that everyone should be able to have inspirational encounters with wildlife, and there's something slightly ironic about the fact that the sight of the hairstreaks isn't available to people more easily, given that they're currently one of our less-threatened butterflies due to the prevalence of oakwoods.

As with other species of mature woodland, though, I wonder what the future holds. The end of coppicing and woodland management has created a monoculture of oaks that will decline and die over a relatively short period, and without some variation in the structure of these woods we won't see a new generation of trees begin to develop. The restoration of coppicing at places like

Spring Park, and subsequent, albeit difficult, decisions about felling the occasional older oak will allow saplings to develop and actually ensure a succession of maturing oaks that will host clouds of hairstreaks long into the future. A renewed market for oak timber and other wood, if it's felled and extracted with care, could provide the economic support for some of the management our ancient woods need. And whether visitors on foot are able to experience the hairstreaks or not, the flood of light and warmth that comes with the restoration of management will support many more butterflies, and the flowers they rely on, in a revived spectacle that is accessible to all.

CHAPTER TEN

Treading Softly

The hidden life of ancient soil

THEY ARRIVE IN an old lorry painted silver and red and with one of its big steel lockers tied shut with baler twine. Daniel Brown, the horseman, follows a stocky cob called Rita through the suburban yard at Spring Park and into the wood. I drift out after them and, despite my best efforts not to romanticise the scene, it's like being in a different place and time. The atmosphere is helped by the fact that it's a western morning; there's no rain as such, just a fine drizzle that looks like mist at a distance but is definitely, seriously wet up close. Rita's a Comtois, Daniel tells me – a rare French breed of working horse – and a glistening brown that, fittingly, can only be described as chestnut. A blond fringe falls jauntily across her eyes as she plods steadily past the tangle of felled timber and bramble while we scout her first load.

Working horses, like coppicing, have been integral to human history in Britain since the Bronze Age, and they've almost certainly been working in the woods for most of that time. In recent years there's been a resurgence of horse-logging as woodlanders seek to restore active management while protecting the woodland soil, an astonishingly complex ecosystem that underpins the ecological significance of ancient woodland. As we strive to improve the condition of our woods, whether by introducing structural diversity or clearing inappropriate conifer plantations, there's a new place for old technologies like horse-logging in getting sensitive work done. Modern woodlanders also need to embrace innovative ideas that can achieve the same things, as well as finding new ways to pay for our essential work in response to the crisis in nature and climate.

Jamie and Jim have done their usual efficient job in felling the overstood stools, leaving us a scattered pile of stems ranging in diameter from the size of a bean can to well over half a metre. This was previously one of Spring Park's remaining areas of unrestored coppice, down in a corner between the office and the Anglo-Saxon woodbank; it's the mix of sweet chestnut and small-leaved lime that I explored with Hugh and Jonathan and which had been neglected since its cutting during the Second World War. We realised that, belying the species' reputation for immortality, some of the old lime coppice is in fact dying. Where stools have blown over, the fallen poles that rest on the ground – and might therefore normally take root – don't produce new foliage because they're too shaded; the elevated poles that are suspended above the soil, on the other hand, spring back to life for a time but have no chance to set roots, then fail as parts of the old stool decay and collapse. Factoring in the wider benefits for nature that restoration would provide, we decided to take the plunge and managed to get a grant to pay for Daniel and his horses to pull the wood out; this is, for us, a novel experiment.

While it's easy to carry the produce of hazel coppice out by hand, the fatter neglected poles of canopy species like chestnut and lime – and even in-cycle chestnut, harvested at around twenty years old – need something with a bit more grunt, which typically now means tractors and machines. After just a few years at Spring Park I've seen a range of mechanised approaches. In my first year a forestry company, desperate to meet their contract to supply woodchip to a biomass power station, cut the chestnut coppice using a tracked harvester that rumbled through the coupe and cut the poles with its terrifying grab-saw, leaving a mess of both the stools and the ground. In the second year, the rangers had cut it with chainsaws, but we'd still driven our big tractors in, towing a chipper and trailers; we left the stools in better nick, but Baz and I felt increasingly uncomfortable with the ruts we left in the woodland soil.

The following year we tried pulling the poles out using a forestry winch mounted on the back of a tractor, the idea being that we wouldn't need to drive the tractor through the coupe so much. It was an improvement, but my work of 'choker-setting' – hooking the winch cable to the timber – finally helped me understand the true richness of the ancient woodland soil, as well as showing me that we still weren't treating it with enough care. Choker-setting has to be the dirtiest, most unpleasant and downright knackering job in the woods. I love it. In theory, it's simple: you fasten the fallen timber to a tractor or winch that will drag it to a track. But with the trunks typically driven into the ground by their weight, you normally have to get down on your hands and knees among the leaf litter and feel around for a little gap beneath the log, or a spot where you can dig the hook through, before snagging it back on to the cable so it tightens under tension, 'choking' the wood.

Like every practical job there's an art to it, so subtle it's hard to explain. Really experienced setters can find invisible soft points under buried logs and shove the hook through quickly where newcomers have been scratting around hopelessly, very literally up to our elbows in mud. The intimacy, though, is part of the joy of it. There are few other jobs in the woods where you get quite so up close and personal, hugging the bark or face down in the earth, scrambling about on all fours; it's like climbing, but in among the soil and bramble. Exertion and extreme proximity accent the beauty in the little features that normally blend into the whole. A thousand shades of brown in the wet leaf litter glisten into red and yellow; bramble rope, green at first glance, crystallises into pink and crimson around the base of the perfect shark-fin thorns; a lonely, late-winter beetle making its purposeful way under the ground vegetation seems to expand in size to dominate the scene for a few moments.

But it was the close focus on the soil that brought home just how special it is. I was used to getting into the earth by this point, having dug plenty of holes for the straining posts of fences or to hang gates. But I realised, with my face in the leaf litter at Spring Park, that what I was in among was something completely different to the soil I'd dug through in fields, most of which had been farmed conventionally in recent years and had been ploughed and fertilised. Out in the farmland, there'd typically been a simple, thin band of organic matter beneath the grass and over the stonier stuff beneath; here, where dragged poles had cut a rough section through the earth, I saw that there was much more going on.

Last autumn's leaf litter, the depth of my hand, sat above a deep layer of crumbly life grading from recognisable scraps of leaf and twig to a moist, cake-like base where the detritus from the trees and plants had been progressively broken down by fungi and been eaten and excreted by soil organisms. This section of the soil, called the humus layer, is much thicker in ancient woodland soils than in any other habitat; the leaf litter and woody debris, the diversity of plant species, and the subsequent richness of fungal species combine to create something that simply doesn't exist in quite the same way outside the woods. Even in late winter there were clues to the rich ecosystem within as I saw fat earthworms snaking through its light structure, woodlice scurrying urgently from a disturbed chunk of decaying wood, and lightning-fast millipedes winding manically around nubs of twig and little pebbles.

Creatures like these, though, are the giants of the soil world. Invisible without a hand lens – or, in many cases, a microscope – is an incomprehensible range and abundance of tiny organisms that make intact temperate soils some of the world's richest ecosystems. Each hectare of woodland soil can hold more than a ton of invertebrates,[1] and, as George Monbiot points out in *Regenesis*, scientists believe that only 10 per cent of the smallest species have even been identified.[2] Among the thousands of species in each metre of healthy soil, the majority can't be found in any book.

Thom Dallimore, a biologist at Liverpool Edge Hill University, describes the soil fauna as a cross between the African savannah and the work of H. R. Geiger, the sci-fi illustrator who designed the creature for the *Alien* movies. As with the invertebrates we met in the decaying wood of ancient trees (itself in the process of returning to soil), vast dramas are played out in miniature among mind-bending ecological niches; there are herds of grazers feeding on detritus and fungi, and whole webs of predators and parasites. When visualised by Thom's extraordinary micrographs – images created using an electron microscope – we see that the invertebrates are fascinatingly fierce and otherworldly. Soil mites are some of the most numerous creatures on the planet and one of the most populous is the oribatid mite, a slow-moving detritivore that's heavily defended by a bulbous shell. It's joined by pseudoscorpions that look just like tiny versions of their namesakes and springtails that do, indeed, have a spring on their tail that they can activate to bounce away from predators. A swift getaway is necessary as some of the hunters operate by injecting chemicals into their prey's carapace before sucking out the resulting goop, others by reeling in their food after snaring it with a shooting, chameleon-like tongue.[3]

Even smaller and more numerous than the mites and springtails, which might themselves be just a tenth of a millimetre in length, are the nematode worms and protozoa. These are single-celled animals that feed on bacteria and there might be many hundreds in every spoonful of woodland earth, far more than in the same amount of typical farmland soil. Along with soil microbes – there could be one billion of these, of 10,000 species, in a teaspoon of healthy soil – these smallest organisms contribute to the soil's essential chemical role. As well as breaking down dead plant matter and making its nutrients and other components available for new life to thrive, healthy soil ecosystems play a vital role in absorbing and regulating atmospheric gases like carbon dioxide and methane. They also influence the balance of other bacteria and the production of hormones that regulate the growth of plants and help them fight disease.

Obviously, little of this was visible to me as I eyeballed the soil while straining to wrap the rough choker cable around the chestnut poles. The humus layer was, though, clearly different to the equivalent layers of soil I'd experienced while digging in fence-posts. As well as being thicker, it was softer and slightly stickier in texture, and it smelled more strongly but also more pleasantly. There was an almost-mouth-watering scent of fungi and wood and decay and plant matter that is, I reflected, unique to woodland soil. The smell is also, I've been told more recently, a good sign of its health, and results too from the chemicals emitted by its rich variety of bacteria: soils in poor health typically smell more sour and less appetising. The open structure and the way chunks of it fell apart in my hands like the moistest of chocolate cakes is an essential component of its quality. The whole range of soil organisms – from the microbes and protozoa through to the worms and woodlice, not to mention plant and tree roots – rely on being able to move through the spaces between the fresher detritus, decomposed organic matter and the sand or pebbles of the mineral layer with which it blends below.

The immeasurable, microscopic variation in the soil's structure provides boundless niches for those tens of thousands of different species. They need oxygen and water in a delicate balance, just like above-ground species, so that they can feed, hunt and reproduce, and for their essential chemical processes to take place. Where the soil is ploughed or crushed by big machinery the gaps are squashed shut and many of the organisms don't have the space or the oxygen to survive, while the change in the delicate balance of gases plays havoc with the chemical reactions – a situation made infinitely worse by the application of artificial fertilisers.

Felicity Roos, the National Trust's adviser on soils, told me that the importance of ancient woodland soils 'lies in the fact that they haven't been disturbed since at least 1600, so they've retained their complex relationships – fungal and microbial – that take centuries to develop.' That long-matured richness – the organic matter that's incorporated into the soil at every stage of decay, the organisms within it, and the tree roots and fungi – is also a huge store of carbon, holding even more, in fact, in many woods, than the trees that grow from it. Physical damage, changes to the chemical composition and wider threats like climate change, which brings both more droughts and more periods of waterlogging, can all affect the soil processes and lead to an increase in the rate at which carbon dioxide is emitted from the soil.

I could also see tiny white filaments winding through the exposed strata of humus that were both dense mats of the thinnest tree roots and fungal hyphae, just-visible threads that form the 'body' of fungi; they grow through the soil and around and inside the plant detritus that some of them consume and break down, as well as into and among the roots of living trees and plants. There are thousands of species of fungi in our ancient woods, forming a core component of their soils, and, in common with the soil organisms, many of them haven't yet been named or categorised. Like the other elements of the soil ecosystem, they play an immense range of ecological roles and are similarly vulnerable to changes in the structure of the soil and its chemical composition. While some of them are integral to the nutrient cycle, breaking down tough wood that most organisms can't consume without fungi processing it first, other species bind plants and trees even more comprehensively into the life of the soil.

The role of these mycorrhizal fungi (the name literally means 'fungus roots') has captured many people's imagination in recent years thanks to scientists and writers like Suzanne Simard and Peter Wohlleben. Mycorrhizae form symbiotic relationships with nearly all plants and trees, taking from the plants carbohydrates produced by photosynthesis and, in return, massively increasing the range of the plants' root systems as well as helping them to access a wider range of nutrients in the soil. Some of these nutrients, such as zinc and copper, are almost impossible for plants to absorb without fungal assistance.

Inspiring even more public interest, however, is research by Simard and others that has shown that, as well as increasing the amount and range of nutrients available to any one individual tree, mycorrhizae can also convey carbohydrates between trees, radically altering the previous assumption that woodland ecosystems are defined by competition and offering the prospect that trees might work in community to share resources.[4] And more than that, researchers have also found that trees might communicate in other ways through their fungal associates: for instance when hormones are triggered by a stress factor in one tree – perhaps the intensive grazing of its leaves by caterpillars – neighbouring trees have been observed modifying their physiology in response to the impending threat, like producing more tannins in their leaves to make them less palatable.[5]

This concept is often known as the Wood Wide Web, a phrase coined by Sir David Read when he introduced Susanne Simard's work in the journal *Nature*; it's an evocative phrase that is helpful in envisaging – and communicating – the interconnectedness of the woodland ecosystem. As Merlin Sheldrake points out in *Entangled Life*, though, there's also something not *quite* right about it.[6] The analogy applies our understanding of a recent human invention – massive and complex as the internet is – to a foundational natural system that's millions of years old and infinitely more sophisticated, and he describes the model as 'plant-centric' in its assumption that fungi are merely the connectors, rather than a central part of the story. Some ecologists also feel a sense of disquiet with the possible over-reach of the current science in the idea popularised by some writers of trees communicating in a sentient, humanised way; others make the point that, while the fungal transfer of carbohydrates between trees has been proven, the actual quantities involved are minimal – and that connectedness does not automatically equate to cooperation, in a human sense, but simply a greater range of mechanisms for each organism to get the things it needs.

Much of this science is recent and developing rapidly. However we characterise these relationships now, it seems likely that as our understanding grows it will only become more clear that woodlands should be viewed as profoundly interdependent communities rather than just groups of trees. It seems safe to assume that we should view plants, fungi and soil – with its infinite fauna and microbial life – as constituting an inter-reliant and incredibly important system rather than a collection of individuals, irrespective of ongoing debate about exactly how it works.

Felicity also explained to me that the richness of ancient woodland soil makes it important as a living experiment, each ancient woodland being a unique and undisturbed island of soil ecology – 400 years old at its youngest – that provides an example of what we might have lost from the surrounding landscape through ploughing, fertilisation and other human modification. As we aspire to a more sustainable or 'regenerative' approach to agriculture, she told me, with a focus on soil health, 'we need to know what we're aiming for. Soils in ancient woods can help us understand what a really healthy soil looks like, and help us work towards it elsewhere.' Among other things, this probably means increasing the diversity of plant species on farms, including establishing more trees. If we get some of the necessary changes to land management right over the next few decades, Felicity hypothesised, 'in a couple of hundred years lots more soil on farms might look like ancient woodland soil'.

With the dawning realisation of just how special the soils of our oldest woods are it became clear that, as enjoyable as I found choker-setting, even this lighter method of extracting the wood with the tractor and winch still isn't right. The tractor had to make so many trips along the same few routes that we churned up ruts anyway; not as bad as previously, but still more than we could accept. It made us realise, though, that the distance to the wider paths that surrounded the coupe isn't really that far, so in the last few years at Spring Park we've taken it right back to basics and carry even the in-cycle chestnut out by hand – what land activist Simon Fairlie has called the 'neglected renewable resource' – stacking it by the brick tracks for collection. We have to choose the right team, as anyone who isn't fully on board hinders more than helps with the amount of moaning they do, but we've found that four of us can easily shift 5-metre lengths the width of dinner plates at their fat end, and the place is pristine when we've finished.

The overstood chestnut and lime in the corner by the woodbank is a different matter. With some stems as thick as a post-box there's no way we could carry them out so, now averse to relying on the big machines, we've brought in Daniel and his horses. Daniel asked us to leave the timber in long lengths so he could judge how much the horses would be able to pull depending on the girth, and I'll balance this with the couple of standard lengths we know we'll be able to sell. Three metres for the fat stuff, five for the rest, but it depends a bit on the species what the horses can manage – lime is much lighter than chestnut – and the fattest logs have to go even shorter. This means that the timber's in much more of a jumble than we'd usually leave it, and my job is to cut the lengths following Daniel's instructions.

Rita stands patiently on the path while we have a quick recce. Daniel clocks the first couple of poles he can pull without any cutting, then points out what length I need to cut a dozen more to, scrambling across the pile and speaking only as much as we need: 'Five... five... three... five...'. While I grab a saw, Rita starts pulling her first log down the slope, placing each hoof precisely but to a quick rhythm. Daniel hangs on to the long reins and skids and bounds after, just to the side of the load, occasionally leaping a particularly bushy bramble patch or casually side-stepping over the timber onto clearer ground. I watch that first one go, grinning, and I'm glad to have something to do alongside them.

Like the tradition of coppicing itself, horses are intrinsic to the story of British woodland, and to human history more widely. Before the development of the train and then the internal combustion engine, they were *the* major form of land transport; there were more than 3 million working horses in Britain at the turn of the twentieth century, in comparison with less than a million horses today – most of which are kept for pleasure.[7] Just as with our lost connection to woodland, I find it fascinating to think about what this would have actually meant for our ancestors in terms of their day-to-day relationships with horses: they would have been involved in almost every facet of life, where for most of us now the only traces left are a cobbled stable yard or mounting steps behind an old pub.

It's thought that the horse was probably domesticated in Britain during the Bronze Age – cartwheels have been discovered among the wooden artefacts at Flag Fen – and by the Iron Age they were in common use for hauling goods as well as for riding and in battle; the first Romans in Britain were reputedly met by an army that included 4,000 chariots. Although we can't know for sure that horses were used in the movement of coppice poles it seems almost impossible that they wouldn't have been, as coppicing was already established, and certainly by the Middle Ages the development of the system of coppice with standards relied on them.

Daniel, along with most modern British horse-loggers, uses smaller breeds of working horse, like Rita and her companion Eddie, of a physical type often called cobs; they can move nimbly over rough ground and around trees and stumps. The cobs are harnessed to a wheeled timber arch or 'neb', today made of metal, that fastens to the horse's harness using rigid poles and lifts the chokered end of the pole off the ground to stop it digging into the earth, as well as ensuring that the wood doesn't slide into the horse if it gathers momentum going downhill. Some horse-loggers harness smaller loads directly to their horses with soft straps when it's safe to do so, especially on rougher ground or where there are tight turns. With the exception of the neb, which until a few decades ago would have been made from wood, this system of extracting wood from a coupe has probably been unchanged for hundreds of years or more. It relies simply on the relationship between horse and human as the logger trots – or leaps and skids, like Daniel – alongside the load with a long rein, although most of their communication with the horse happens verbally: traditionally 'gee' for right and 'haw' for left.

Along with so many other aspects of woodland culture, by the 1980s horse-logging had almost died out. After the Second World War tractors, bulldozers, and then purpose-built harvesters and forwarders made mechanised felling and extraction the obvious choice for industrial forestry. The increased profile of ancient woodland and the need for its conservation sparked a renaissance and today there are a couple of dozen horse-loggers at work in Britain, although some of them, like Daniel, also do lots of other horse work. Those that specialise solely in logging typically end up travelling far and wide to stay busy, living in their horseboxes in the woods for weeks on end.

'Do I need to keep my distance from the horses with the saw?' I ask Daniel. He sounds almost offended when he answers: 'Come as close as you need.' Every now and then he checks my progress and we stop for another business-like session of agreeing lengths. Initially, Eddie stays at the horsebox with Daniel's mate Dave, who's along for the ride. It helps Rita pull harder, he tells me later, because she wants to get back to Eddie. He's a perfect piebald cob, the working horse of my dreams, but young and inexperienced. Once Rita's in the swing of things Dave starts bringing Eddie up to the wood too, and, a little more slowly, and with a bit more care and attention, he starts to pull some poles out. It's immediately obvious that Eddie loves the work. He seems almost giddy and fidgets much more than Rita while he's waiting for Dave to set the choker, and always sets off too soon when he thinks it's ready, so Dave has to hold him back with a big laugh. Dave isn't far off being as strong as Eddie anyway; he lifts big chunks of wood to pass the chain underneath with his rollie hanging from the corner of his mouth, chuckling self-consciously when I talk to him.

After a couple of hours I'm done cutting. Daniel glances up from where he's setting a choker around the third of a group of slim lime logs and smiles: 'You've been having fun.' We're all soaked and the mud has spread in a gradient, from our caked feet and knees up to mere grime on our faces. With the choker set he clips the chains into the yoke, picks up the reins, and Rita – who's just been stood there, steaming gently – is off with barely a sound from Daniel, knowing exactly what to do. I follow them along the path, round the corner of the field and into the yard, where at least a couple of dozen logs are already laid out in line. Rita takes a big loop round the back of them and Daniel eases her to a stop as they pull alongside the stack. They never seem in a rush. 'You've got to be calm around the horses,' Daniel says, which is both obvious and incredibly satisfying. His movements to set and remove the chokers are deliberate, almost casual. 'Steady away,' as Jamie says: keep moving, keep working, don't stop to chat. Things get done.

With my bit of the job finished, I can watch them properly as they get back to work. Daniel thinks it's a good time to try some of the biggest logs now the horses – and men – are warmed up and before they start to tire. He produces a couple more chains from a locker in the horsebox, lines the horses up in front of a bit of chestnut over half a metre wide, and fastens Rita behind

Eddie and then on to the wheeled timber arch. There's a bit of checking and muttering between Daniel and Dave; this is evidently not a common pull. The horses seem to know it too and appear just a little less anxious than usual to get started, but Daniel clicks his tongue and they lean into it and get moving, the men almost strolling and the horses noticeably straining on, heads down and shoulders forward. Even more than earlier, they look like they've stepped straight out of one of my old car boot sale woodcuts as the team round the corner and drop the big lump of chestnut by the stack.

They go back for a couple more bigger logs while the horses are working well together, and I check the ground. You'd barely notice they'd been there up among the cut stools and bramble on the slope, and, while the path along the bottom is slick with mud, it's a slippery surface churn rather than the deep ruts left by tractors; it'll recover in no time. They've done forty logs so far, probably half of what a tractor forwarder might have done – but then we'd have had to use a winch to pull the timber down to the path first. So it's not too far off, and more importantly there's no damage to the ancient woodland soil, with its fungi and invertebrates and everything else. Back at the edge of the wood a little crowd of dog-walkers has gathered and everyone's chatting away, hopelessly enamoured with Rita and Eddie. They're interested and excited about our conservation project, an additional benefit you don't get with tractors smashing around.

While my experience of seeing machinery used inappropriately to manage coppice is far from unique, perhaps the biggest threat to our ancient woodland soils is during the restoration of Plantations on Ancient Woodland Sites, the PAWS that, according to the Ancient Woodland Inventory, constitute nearly 40 per cent of our ancient woods. Forestry harvesters are often – perhaps normally – used to clear the conifers that were planted in old woods after the Second World War, partly because it's the most efficient way to do it, but also, I think, through an occasional lack of understanding and an assumption that we can approach them like we would any other conifer plantation. But there's a particular tension to the mechanised restoration of PAWS, in that their ancient soils are the key remnant feature from which a restored ecosystem will emerge – so heavy-handed work risks damaging the very thing we're looking to conserve.

The dense ranks of conifer that were planted after the broadleaved trees were cleared, and their stumps and stools bulldozed out or poisoned, don't host anything like the wildlife that the native trees they replaced were home to. The non-native trees simply don't support as many species as the oak, beech or ash that were torn out, and neither do most post-war conifer plantations have the varied structure and changing areas of light and warmth that much woodland life needs, so the uniform rows of trees shade out the woodland ground flora and don't provide for many insects or birds. The seed bank typically persists in the soil, though, along with much of the rest of the special soil ecosystem, forming an important remnant of the wood's former value – although the chemical composition of the soil, and the life it supports, can also start to change for the worse with the influence of acidic conifer needles. It's not uncommon to find some of the more shade-tolerant plants, like hart's tongue fern, clinging on among the woodbanks and charcoal hearths that we might also find hidden beneath the spruce or hemlock. Once the conifers are removed many of the flowers spring back to life from their dormant seeds, while the wider woodland ecosystem recolonises a resurgent layer of native shrubs and trees.

Removing these inappropriate twentieth-century plantations is therefore one of the most important things we can do to restore the value of ancient woods for wildlife, especially given the scale of the threat identified during the inventory surveys. PAWS restoration is a quietly developing success story in Forestry Commission woods and among the conservation charities like the Woodland Trust and National Trust; between 2010 and 2018 the long haul of restoration began in around 27,000 hectares of over-planted ancient woodland in England. While this might not sound like a huge proportion of the total PAWS, which is estimated at around 150,000 hectares, it's still a lot of woodland that's receiving difficult and time-consuming – but very positive – management.[8]

In places like the Teign Valley on the eastern fringe of Dartmoor, the Woodland Trust and the National Trust have been working together to sensitively clear conifers like Douglas fir, spruce and western hemlock from temperate rainforest that is now responding with flushes of woodland wildflowers and a resurgence of native tree species. I made an inspirational visit to the valley to see how things were going during a big grant-funded project, and was amazed that it took us a full day to make our way along its 5-kilometre length; we couldn't drive more than a few minutes in the rangers' Land Rover without stopping to speak with different work parties, the wood bustling with people all deeply involved with restoring its health. As well as a couple of sensitive contractors hand-cutting tall larch that was then dragged down the hill by long-armed excavators parked on the track, we met rangers finishing a coppice restoration job among the oaks, local volunteers fixing a deer fence, engineers improving tracks and planters fixing protective tubes around new native trees in an area of PAWS cut the previous year. At the heart of it all there's a little sawmill installed by the Woodland Trust, where local woodlanders were turning some of the wood produced by the conservation work into materials needed by the two charities and other land managers: gate- and fence-posts, planting stakes, weatherboards and more.

Conifers were planted across many ancient wood pastures, too, as I've seen on other trips. At Croft Castle countryside manager Iain Carter oversaw the clearance of hectares of western hemlock, one of the shadiest trees, revealing in the process dozens of extraordinary ancient oaks. Fifteen years on from the clearance, following conservation grazing by hardy cattle, the landscape looks as close to a working medieval common as we're likely to find today; each spring it's a sea of bluebells around thickets of bramble that shelter young hawthorns and oaks. Jamie Simpson spends a week there every winter cutting young trees into new pollards, reinvigorating the old tradition with a new generation of working trees, and I've come to know the place having helped him out on a few occasions. Like Brian Williamson and Dave Watson with their relearned specifics of coppice-crafts, Jamie has taught himself how to successfully create pollards by close observation and trial and error, and is now easily the most expert practitioner of this ancient art in Britain.

The Forestry Commission agreed to bring their neighbouring plantations at Croft into a similar scheme, spreading the benefit for wildlife across what has become an extensive wood pasture landscape. They revealed in the process a fascinating, if depressing, aspect of the twentieth-century process of 'coniferisation'. Close to the entrance from the road, the removal of the non-native trees revealed the stumps of ancient oaks that had been felled when the plantation was made, an intimidating prospect even with modern chainsaws. Further in to the site, and presumably later in the planting project, there are dead hulks of oak that were 'ring-barked', killed standing by having their cambium – the living wood directly beneath the bark – removed in a strip around the base; further in still and some of the oaks have clung to life, the ring-barking having been done less diligently. Finally, at the most remote point of the plantation, there are a handful of untouched old veterans that have now been carefully opened up and are slowly responding with new life. We assume that the work of killing and removing the big old trees was simply so hard and time-consuming that the energy of the foresters tasked with the job fizzled out, which preserved those last few special trees.

Given that the ultimate importance of PAWS lies in the soil, though, it's ironic and frustrating that efforts to restore these special remnants of ancient woodland can sometimes damage its most significant surviving feature. Over the past couple of years, woodland expert Ian Rotherham has raised the alarm around the inappropriate use of heavy machinery in ancient woods, particularly during the restoration of PAWS. He has gone so far as to describe it as 'The New Locust Years', riffing on Oliver Rackham's characterisation of the post-war period, and cites many examples of ancient woods where heavy forestry harvesters have been used to pluck out conifers but have damaged the woodland soil and its archaeology.

The fact that the two conservation charities relied on a one-off grant to get so much restoration work done in the Teign Valley exposes the challenging economics involved in restoring PAWS and helps explain why some operations have been less than sympathetic. Although many ancient woods are on the steeper ground and wetter soils that were least attractive for farming, it's possible to get big machines on site in many places, and the income from the large quantities of softwood timber produced can then often pay for the work. There's sometimes simply an assumption that, as we're removing conifer plantations, we can just do it in the way we'd harvest a normal conifer crop that isn't sitting on such important and vulnerable soil, but this can lead to devastating rutting and compaction.

Economic pressures can influence the speed of work, too, which can also have an impact on the soil. It's typically preferable to remove the conifers gradually, opening up the canopy over a number of years or even decades, protecting the soil from erosion and creating optimal conditions for new native trees and woodland wildflowers to develop. This has the additional advantage of reducing the amount of vehicle movement in any one operation but, much like using smaller machinery, it's often harder to make progressive thinning pay for itself compared to felling everything in one go.

There are occasions when the nature of the soils or the timing of operations means machines *can* be employed with care, and some woodlanders have found ways to make their use more sensitive. At the Ashridge Estate, for instance, the rangers laid festival-style tracking to protect the soil from the impact of a big forwarder making its repeated trips to haul wood out to the road. But in most places we need to find better ways of extracting timber sensitively from our ancient woods, or different ways to pay for good management.

As well as horse-logging, smaller-scale mechanical extraction can work well – and break even – in ancient woodland if it's managed carefully and in the right places. At Forge Wood, Jamie pulls all his wood out using a tractor but does it in summer when the clay soil is firm, and times it carefully around breeding birds. In the Lakes, the National Trust foresters have a mini-tractor and forwarder that's equivalent to horses, in the lower amount of wood it can extract each trip but also in its minimal impact on the ground. (And, as Martin Thwaites, the Trust's head forester in the Lakes, told me, 'It doesn't get tired'.) Even a winch with a longer cable can do a more sensitive job than we were able to in the chestnut at Spring Park, and I suspect that new technology will lead to the development of more lightweight machinery; the rise of electric cargo bikes around hilly Sheffield, for instance, makes me think that similar small but powerful electric motors could drive lightweight forwarding rigs.

I feel confident, though, having seen their deep ruts and churned soil in a number of ancient woods – including routes cut through medieval woodbanks – that the use of a forestry harvester is nearly always a step too far. PAWS and coppice restoration should begin with hand-cutting by careful workers with chainsaws, even if they need some sensitive mechanised help to get the wood out. Wendell Berry, who manages his land in Kentucky using working horses, advocates a return to a form of farming where we recognise that 'what is good for the world is good for us. And that requires that we make the effort to *know* the world and to learn what is good for it. We must learn to cooperate in its processes, and to yield to its limits.'9 The same could easily be said of our ancient woods.

That said, it's essential to remember that hand-cutting trees is a dangerous business. Big machines are vital in making many other types of treework safer, and in those situations it's absolutely right that we default to their use, whether they're harvesters for forestry jobs and roadside safety work or platforms that help us avoid climbing. I don't have all the answers as to how we balance the risks of physical work with the need to protect woodland ecosystems, especially when their decline is also bad for our health – as is sitting still all day, for that matter. But it should go without saying that any woodland work needs careful planning with all these factors in mind, and that everyone involved needs to be well trained and have the right experience.

There aren't any easy solutions to the extraction of timber, and the techniques used will be different in nearly every wood – but we need to start planning our management by understanding the limits of the land. Our work should be defined by the principle that ancient woodlands are so important for their wildlife, soils and the other benefits they provide, not to mention their cultural value, that we need to find ways to manage them with a focus on their health even where the value of the wood produced won't cover those costs. And we shouldn't carry out harvesting operations in a way that does make money if that results in damage to the soil, woodland archaeology, or anything else.

With heavy-handed extraction being often driven by economics, it's vital that we find new ways to make PAWS restoration and the active management of ancient woodland pay. There is potential, as we saw among Spring Park's oak standards, for the value in sustainably harvested wood to rise, whether through a renewed tradition of growing high-quality broadleaved timber or following the development of exciting new technologies that create engineered products from less uniform wood. While these opportunities could make good management more viable for conservation-minded woodlanders, or enable the use of more sustainable extraction techniques, they also bring a risk that they might drive less enlightened landowners to cash in – so improved regulation will also be important, to ensure woods are managed with a focus on their many benefits to all of us.

Markets for some of the less tangible products of woodland are also developing, with landowners increasingly able to take payments for the 'ecosystem services' provided to society by their land. Savvy landowners can already sell the carbon their woodland stores, monetise its ability to slow the flow of water and mitigate the risk of flooding, and, with the launch of 'Biodiversity Net Gain' in England, even sell its ability to support nature through a mechanism where developers have to pay to create or sustain habitats if their work damages others.

In my professional life I'm a pragmatist, and it's clear to me that as long as the current economic system prevails the owners of ancient woodland will need to engage with this approach, taking payments for the provision of ecosystem services to fund the restoration and good management that our woods desperately require. On a personal level, though, I have to admit that I'm deeply uncomfortable with the marketisation of these fundamental natural systems that underpin life on our planet. I saw a job advertised that involved 'scaling investment in nature assets' by creating 'frameworks to facilitate transactions between project developers and investors/corporates'. Is this really the way we redefine our broken relationship with the planet and its ecosystems? As we've seen too often during the last couple of centuries, if you put a price on something someone might be willing to pay it; and besides, it's a recent development that price is the only way to confer value. Economist Guy Standing points out that to refer to 'natural capital', as approaches like this are sometimes described, is to commodify nature; not only is this bad for nature, it's also a form of enclosure that converts this aspect of the modern commons – a fundamental resource in which we all have a stake – into a source of profit for individual landowners.[10] I would rather we viewed a truly sustainable approach to water, soil, wildlife and other so-called resources as a fundamental red line above which a different sort of economy must operate; one that sees the market yielding, as Wendell Berry wrote, to the limits of the land.

The best way we can help landowners do this in the short term, in my view, is through stronger regulation combined with government grant payments that genuinely cover the costs of good active management, done in such a way that all the important features of a wood – including the soil – are protected. Landowners don't always like more rules and oversight, but we could see this more positively as the creation of a new and more sustainable contract with the natural world on which we rely, a societal agreement that existed throughout most of history under commoning and the customary practices that pre-dated it.

There's already a package of government grants for woodland management in England, but it's OK at best; the payments typically won't stretch to the costs of good active management in woods with poor access, for instance, where there's no chance of extracting any wood for sale, or for the loss-making restoration of coppice. There's scope to improve the grant regime to ensure more ancient woods receive the sensitive management they need. Although the grants are payments for ecosystem services, the people stumping up the cash are all of us, through our taxes, rather than corporations offsetting detrimental impacts elsewhere. As well as linking the beneficiaries of good management – that is, everyone – with the few people in the fortunate position of being able to provide it, adequate grants for the exemplary management of our ancient woods would also simply be a powerful statement of how much we value them as a society, and a recognition of their significance to the past, present and future of British life.

There is also, of course, an important role for the landowning charities like the National Trust, Woodland Trust, Wildlife Trusts and RSPB in managing their ancient woods to a high standard, which they do in many places. The simple fact remains, however, that they can't afford to look after all their woods to the level we need – and, between them, they own less than 15 per cent of Britain's broadleaved woodland.

When I turn up at Spring Park the next morning, Dave is busying around frying eggs and brewing up in the wagon, a timeless moment in our suburban yard. Outside Daniel brushes Rita, murmuring to her. Eddie fidgets close by, ready to go again.

As a woodland history nerd, I'd be lying if I said that working with Daniel, Dave, Rita and Eddie isn't incredibly exciting. But, although the job doesn't break even when we sell the chestnut and lime they pull out, it's not simply an exercise in nostalgia, and using the horses is still the right thing to do; apart from the bramble looking a bit battered in the coupe, you can barely see they've been here. The precious, delicate soil ecosystem, with its astounding abundance and complexity of hidden life – not to mention deep stocks of carbon – is almost entirely undisturbed, while the lime-stools spring back with their masses of soft new stems, surrounded by wood anemone, celandine and stitchwort. Tiffany Francis-Baker asks, 'Who else but the horse can reflect both the tamed and untamed elements of our world, bound together in one warm, mortal body?' It is perhaps no surprise that horses work so well in the care of our ancient woods, themselves places that are neither purely wild nor purely human but, at their best, a melding of both for mutual benefit.

Daniel switches things round again as we get started. Eddie was feeling a bit hemmed in working in a team, he thinks, and was chucking his weight around, starting too slow or going too fast – he prefers the freedom of pulling his own logs. 'Rita don't care either way.' The length of each trip is growing as they clear the coupe, and with it comes some trickier terrain. The horses don't blink at the steeper slopes and rough going, but the men have to leap even more dramatically down the little inclines or around groups of saplings. They finish, though, seeming as fresh as they started: steady away.

CHAPTER ELEVEN

Wilding the Woods

Ancient woodland and rewilding

I HELD AN oak leaf the size of both my hands together and felt dwarfed by my outlandish surroundings; this was a familiar space but on an entirely unfamiliar scale, the forest full of tree species I'm used to seeing at home but each half as big again. It stretched away for tens of kilometres and we were under strict instructions not to wander off, but even with the doomful threat of being shot by twitchy border guards it was all we could do to stay in sight of one another as everyone dashed to inspect another novelty or lagged behind to photograph a particularly exciting find.

The psychedelic effect created by the size of the trees and exuberant tangles of fallen stems was enhanced by us having arrived in Białowieża at the height of autumn. The space beneath the canopy – there must be a word for this in some language – glowed with a soft-focus light as the changing hornbeam leaves captured and redistributed the sun's rays. There were huge oaks and maples just on the turn, occasional dark spruce, and my favourite, small-leaved lime, their delicate sprays of heart-shaped leaves still a lush green. There were aspen bigger than I ever thought they could be competing for room in the canopy, their falling leaves hypnotic as they drifted past and spread endlessly beneath our feet, each one a pristine block of rich colour.

Białowieża, on the border between Poland and Belarus, is famous as the biggest surviving remnant of supposed 'primeval' woodland in Europe, although the reality is more complicated. Large areas at its core have always been wooded and never cleared for agriculture, and some of that hasn't been cut for underwood or timber since at least the Middle Ages – but the actions of people have influenced it more than the popular myth would imply, as we'll see. It covers 3,000 square kilometres and about half of it forms a specially protected area that's designated as a UNESCO World Heritage Site. The protected area continues, in theory, to receive no human intervention, so the evolution of the landscape is meant to be subject only to natural processes, and access is by permit only. Through a jammy series of events begun under a tree – where else? – at the Ancient Tree Forum conference at Knepp, Jamie and I, along with my fellow ranger Zuza, were invited on an EU-funded visit to learn about decay fungi with specialists from the University of Helsinki and the Polish forest authority.

The size and antiquity of the forest, with its associated lack of clearance or even much disturbance, gives a rare insight into this aspect of the natural state of European woods before their fungal ecosystems were impoverished by fragmentation, conversion to agriculture and management – even traditional management – that denuded them of decaying wood and depleted their soils. There were fallen hulks throughout the wood; the fungi that break them down are richer in Białowieża than anywhere else in Europe. Every collapsed tree was peppered with a range of fungal brackets, as were many of the living trees. There were bulging semicircular shelves, some hard and woody, some soft and rotting, and shaggy mounds like clumps of miniature stalactites. They varied in colour from bright, wet white through every tone to matt black, and in places there were so many of them that it felt like some sort of art installation, another psychedelic twist on our home turf. A miniature forest of mosses, ferns, flower-less woodland herbs and tiny saplings blanketed the top of each prone trunk; so intense was the fungal activity and its integral role in the ecosystem that it felt as if the whole lot was being consumed back into the soil almost in front of our eyes.

Our trip was inspirational, revealing the richness of 'old-growth' life possible in the soils and decaying wood of a landscape at such scale that's always been wooded; there may be twice as many species of some groups in Białowieża as in British ancient woods, including both plants and fungi. It was also simply a thrilling experience to spend time in such a giant woodland that's home to the big, wild, ecosystem-driving animals that are missing from Britain: bison, wild boar, wolves, lynx and beavers.

The forest felt like the wildest place I'd ever been – certainly in Europe – and the trip helped me understand the exciting potential for the reintroduction of species like these to British woods, one aspect of the influential rewilding movement. But we also came to realise that despite its great age Białowieża doesn't genuinely replicate the wildwood, as was once believed. 'Wildness' can be born of time – that is, time without intensive human pressure – but ecosystems only function properly with the right ongoing influences and processes driving the dynamic, kaleidoscopic effect that was inherent to the wildwood. Although its age and scale have allowed some groups, like fungi, to survive, Białowieża has become unnaturally dense and dark because of the decline in wild herbivores, and, like so many ancient woods, doesn't support as much wildlife of other groups as it would if there were more shifting areas of light and warmth. I learned, for instance, that, although there are more species of bird in the forest than in even the most well-managed British woods, there are fewer individuals per hectare. The monocultural gloom of the woodland was evident to us when we visited; we saw few birds, and large parts of the looming space beneath the canopy, although atmospheric and beautiful, were devoid of scrub species and saplings.

To understand the true potential for wilder woodland, I've had to also look closer to home, at rewilding projects like Knepp, Wild Ennerdale in the Lakes and Wild Exmoor in Devon and Somerset. Despite their young age and the fact that their starting point was typically ecologically impoverished, conventional farmland – its plant species, soils and fungi having been denuded by pesticides, fertiliser or overgrazing – these projects have seen a rapid increase in wildlife due to their sensitive use of grazing animals and the subsequent constant transformation of their progressive woody growth. Even though ancient woodland has a long human history and holds such high cultural significance, it will play a vital role in projects like these, acting as a source of biodiversity for the surrounding land while itself becoming healthier under the influence of low numbers of herbivores. Our ancient woods can bring the depth of time – the evolutionary memory of intact ecosystems – to amplify the power of the more natural, pre-agricultural processes that have been restored to these landscapes.

In many other places, though, our small and fragmented woods, which form the bulk of British ancient woodland, continue to need careful human management: the woodlander is another endangered species whose loss from our landscape has affected countless others. Like Białowieża, our woods contain the irreplicable quality of time in their continuity of habitat, but are also similarly missing the natural processes driven originally by wild herbivores and provided over recent millennia by human management. A resurgent wood culture remains essential to restore dynamism and life to most of our ancient woods, both wild and human; but the new generation of woodlanders can also take the lessons of rewilding to ensure that even carefully tended woods replicate natural ecosystems as closely as possible, providing the best possible homes for their full range of wildlife.

The stillness in Białowieża, despite its astonishing fungal ecology, felt in stark contrast to the abundance we'd experienced shortly before at Knepp, with nightingales singing, other woodland birds resurgent in the developing scrub, and purple emperors swooping into sunlit glades from south-east England's own big oaks. As I saw when I first visited Knepp while working at Ashtead Common, innovative and exciting projects like this can play a different but complementary role to the primeval woods of mainland Europe in helping us understand the dynamics of pre-agricultural ecosystems, as their complex landscapes develop from nothing under the influence of big herbivores and other keystone species.

Knepp has become famous for the conversion of thousands of hectares of arable land to a more natural, 'wilder' approach, but some of the estate's ancient woods have also been incorporated into the large grazing areas, their fences removed and the estate's cattle, ponies, deer and pigs allowed to roam through them. In the decade or more since Ted and Jamie first took me there, I've seen something similar begin to happen in landscapes like Ennerdale in the Lakes, and at the northern edge of Exmoor in Somerset and Devon.

It's relatively early days for all these projects on the scale of tree time, but, where the grazing pressure is low enough, the animals don't denude the ancient woods of their scrub and wildflowers, as can happen when too many sheep get into a woodland, while their partial browsing of understorey species and some saplings should slowly start to open the canopy and help the development of a more varied structure in constant change. Kev Davies, the countryside manager with the National Trust in north Devon, told me that incorporating ancient woods into their landscape-scale wilding project will see 'lots more trees on what used to be fields, and maybe fewer trees in the woods

– where there'll be more open space and disturbance, with all the life that brings. And it will enable constant dynamism and change across the whole landscape, in the woods and beyond.'

Crucially, the ancient woods also act like reservoirs of biodiversity, as Isabella Tree and Charlie Burrell write in *The Book of Wilding*, their long-matured ecosystems overflowing once the agricultural pressure is taken off the surrounding land.[1] Already you can find some of the richest and most diverse scrub thickets spreading quickly out from the edges of the woodland in many of these places, breaking down the artificial boundaries between wood and field while, unseen below ground, their mycorrhizal fungi and all the other inhabitants of the ancient woodland soils are likely to be doing something similar and repopulating the earth that's been depleted by industrial agriculture. Rachel Oakley, project officer at Wild Ennerdale, explained to me that even the impact of an ancient wood adjacent to the area being wilded is dramatic: 'With the intensive grazing of sheep removed, oak, birch and rowan have spilled out into the areas the cows have disturbed – groves of trees are springing up across the open fell behind the wood.' Tree and Burrell write that, at Knepp, 'the ground flora normally associated with woodland... are now marching into the open areas surrounding the woods'. Given how slowly some of these flowers spread, this process, on human timescales, might look more like a trickle than a flood – but it's still something to celebrate.

Grazing and browsing animals lie at the heart of the big British rewilding projects; hardy, old breeds of cattle act as proxies for wild herbivores, sometimes alongside pigs, ponies and deer. The livestock are kept at densities low enough that the theory proposed by Dutch ecologist Frans Vera can play out, with scrub developing where they disturb the ground, while they go on to maintain open areas and other habitat features among the resurgent woody landscape. Vibrant wood pasture develops, so rich as to be revelatory. The term 'rewilding' itself describes a wide spectrum of approaches, though, and at its most extreme – and perhaps most exciting – it goes beyond this herbivore-driven approach to also encompass the reintroduction of charismatic keystone species that have been lost from the British countryside, like beavers, lynx and even bison, each of which add additional complexity and life to developing ecosystems.

Białowieża is known for its herd of European bison, which, along with the place itself, play a powerful role in Polish national identity. They formed a stronghold of a few hundred animals during the bison's extermination across the rest of the Continent, from before the Middle Ages to the Victorian era. The bison were, though, hunted to extinction in the wild here during the First World War, while the last wild bison in Europe was shot in the Caucasian Mountains in 1927. A few dozen survived in zoos and their offspring were reintroduced into Białowieża in small numbers in the 1950s after an eventful period of captive breeding during the 1930s and 40s that briefly involved Hermann Göring acting as their patron. This hiatus in their presence, though, allowed generations of new trees to develop in the forest so that it became the dense woodland that we explored on our visit. It also seems likely that, although the forest was home to Europe's biggest remnant herd, the relatively low numbers of bison for hundreds of years before their extinction meant that this process had been underway for a similarly long time.

The reintroduced bison have bred successfully since the 1950s and there are now around 1,000 in the forest; we didn't see any, but it was exciting to find their hoof prints – longer than my size ten boot – and wiry hairs caught on thorns, and to drop to our knees to inspect their pats, busy with dozens of dung beetles. While the bison should, in time, maintain more glades when trees fall and begin to drive the restoration of a healthier wood pasture landscape, it will take many decades, and probably a bigger herd.

In lots of other places on the Continent the European bison is a prime candidate for reintroduction, both for its inherent value as a rare native species and also as the least-domesticated, biggest and most impactful remaining herbivore, one that can have a transformative influence on its ecosystem. With bulls weighing up to 1,000 kilos, bison crash trees and scrub around as well as browsing them, bringing additional structural diversity and creating all sorts of ecological niches. By dust-bathing, they also create muddy areas and bare earth that can allow willow and other seed to set and grow, and where solitary bees can burrow their nests. There are now around 6,000 bison in wilder-landscape projects across Europe, one of the most high-profile success stories of the international rewilding movement.

There's no evidence for the European bison ever having been native to Britain, although that doesn't necessarily confirm its absence. The wood bison, now extinct everywhere, was here prior to the last Ice Age but probably not since the glaciers retreated. In the West Blean Woods, though, Kent Wildlife Trust and the Wildwood Trust have introduced European bison to a large area of ancient woodland along with ponies, cattle and pigs. Rather than being billed as a reintroduction, the bison simply form the best proxy for the aurochs, the extinct giant wild cow whose effect is all but impossible to replicate with the smaller hardy breeds around today, and which is therefore a missing component of natural ecosystem dynamics. I spoke with Helen Pitman and Matt Hayes from the Wildlife Trust, who told me that the area that's subject to wilding covers hundreds of hectares and is mainly either plantations of conifer in ancient woodland – PAWS – or former sweet chestnut coppice (the bison aren't allowed in the whole area, but the other animals roam the lot). To speed up the process of restoring a more natural landscape, the Trust has undertaken an intensive initial phase of work, heavily thinning the conifer and coppicing the chestnut. Matt explained that these early interventions to open up the closed canopy are often needed to 'kick-start' rewilding projects: 'the woods were nowhere near natural after centuries of human modification, so we had to do a lot before we could get the animals in and let things develop on their own'.

Having let bison and other animals loose in the wood, the team at the Blean have a programme of monitoring underway to assess the impact of this radical approach to managing ancient woodland. Although it's too early to definitively measure its success, Matt feels sure that wildlife is responding to the more natural dynamics: 'The areas of the Blean that are in coppice management are good for wildlife, but they're missing the disturbance and randomness that the big animals bring; as well as opening things up, the animals leave lots more broken trees and decaying wood, and create huge variety in the glades and scrub – it's messy and chaotic in a very good way. Here we're seeing microhabitats develop that we had no idea about, whether they're in the animals' dung, in the different types of churned ground, or simply in the way seeds are dispersed.'

'And let's be honest,' added Helen, 'what we've been doing so far in conservation isn't working; our woodland species are still in decline. We need to give nature a hand so it can get on with looking after itself at a much bigger scale.'

Elsewhere in Britain, rewilding-type projects will inevitably continue to rely on the use of hardy breeds of cattle as the best possible replacement for ancient herbivores and as the main drivers of more natural processes. Cattle alone, however, won't unlock the full potential of our landscapes as we look to replicate the richness of the wildwood. Red deer, ponies, beavers and pigs – performing the role of wild boar – all act as keystone species to one degree or another, driving the shape and function of the fast-developing ecosystems, and creating the conditions for all sorts of other species to thrive.

On my trip to Poland, we were gawping at the bright patchwork of fallen leaves beneath an aspen when we heard the screech of a jay and, although it's a common enough sound in British woods, Renata – our guide – quickly put her finger to her lips. Jays often hang around wild boar, she told us, taking advantage of the rootled earth to forage for bugs. In return, the jays warn the boar of threats; in this case, us. We tiptoed in the direction of some faint – perhaps imagined – rustling, following a definite musky smell, but, alerted by the jay, the boar, if they existed, had disappeared. Whether they were close by on that occasion or not, it was still a heady sensation to know that there was a chance we might come across boar, or bison, or even find signs of wolves or lynx, as well as a reminder of some of the connections and interactions that we've lost from Britain's ecosystems (robins, too, evolved their familiar habit of accompanying digging gardeners by following boar).

And more tangibly, there was evidence of boar everywhere, in big patches of turned earth. Given the increasing importance woodlanders in Britain place on the protection of the soil, it was perplexing at first to see what we might call damage caused by a natural component of the ecosystem. Renata was keen to show us, though, that although the impact of their digging could look dramatic it only extended a few centimetres below the leaf litter; nothing like the crushing, suffocating effect of machinery much deeper into the humus. She also showed us the beneficial role played by the boar's disturbance, where an area rooted by the animals some time ago, still identifiable from its tiny landscape of lumps and divots beneath fresh leaf-fall, was alive with a little forest of saplings. While the boar eat some of the bulbs of woodland plants, Renata told us that the ones they miss do better in their wake with the coarse species of vegetation like bramble knocked back; and the scruffy patches are often transformed by flowers come spring. Ants, meanwhile, need broken ground just like this to begin forming anthills, another fundamental ecological connection that we've lost on our impoverished island. Back at Knepp more recently, Rina Quinlan told a group of us that pigs are the 'key disruptor', an essential component of the project's success, as they break up the grassy sward and enable its initial colonisation by scrub and trees.

Although it's illegal to release boar in Britain – hence the use of Tamworth and Mangalitsa pigs to mimic some of their effects – they have returned to some of our ancient woods. Boar were hunted to extinction here in the Middle Ages, but in the 1980s some farmers started breeding them for their meat using animals imported from the Continent, where they remain populous. Thanks to their abilities as expert diggers and general wreckers, the reason their release is banned, some of the farmed boar escaped and now persist in a few feral populations; it's thought that there are around 3,000 boar living wild in Britain today, with their stronghold in the Forest of Dean.[2]

Their presence remains divisive, to say the least. Boar don't restrict their rootling to the woods, and their ploughing of carefully tended lawns and village greens understandably infuriates some people, while others feel intimidated by their occasional aggressive behaviour to walkers and dogs. Some ecologists worry about their impact on woodland, especially where once-iconic banks of bluebells have been broken up by their digging, degrading their visual impact. Others, however, welcome that disturbance as it diversifies the ground flora and restores lost dynamism to the woodland, as I saw in Poland, as well as simply believing it's the right thing to do to restore native species for the animals' own inherent worth. It's probably obvious that I'm in this latter camp – it was humans that drove boar to extinction in Britain in the first place, after all – and I believe that we can manage the issues boar pose to people as we find new ways to restore our lost connection to nature and rediscover our ability to live alongside other species, as we did throughout history. In the absence of any natural predators, the numbers of boar are controlled by shooting in most of the places they're currently found in Britain. Happily, there doesn't seem to be an overwhelming appetite for their eradication, especially since the concept of rewilding has generated so much interest, and it seems likely that they'll continue to slowly spread to other British woods.

The politics of reintroducing anything that might predate boar, like wolves and lynx, are even more fraught with difficulty; a consultation on the potential reintroduction of lynx to the north-east of England in 2022 ended in acrimony between conservationists, even before the farmers who feared their sheep being attacked had their say. As I write, Northumberland Wildlife Trust is beginning a much better-considered project to gauge the views of local people about a possible release. The return of wolves to the wild in Britain is barely discussed in any realistic way, so emotive is their perceived threat to people and livestock, although both wolves and lynx would almost certainly support the restoration of ancient woodland ecosystems, and healthier landscapes more widely, by preying on deer – one of the chief threats to our woods – and otherwise influencing their behaviour. Wolves, in particular, are naturally re-establishing themselves across much of Europe and, while the process of learning to live with them isn't always easy, people have developed new ways to protect livestock and to compensate farmers for any losses.

There is one predator whose reintroduction doesn't create the same amount of controversy and that nearly all woodlanders, at least, agree on: the pine marten. This beautiful, reclusive mustelid, a cat-sized relative of weasels and stoats but with a deeper red coat, has been driven back to the Highlands of Scotland, with small remnant populations in northern England and North Wales. As well as simply being missing components of intact native woodland ecosystems and worth conserving for their own sake, there's evidence that pine martens might reduce the number of grey squirrels in woods where they're active, both by predation and simply by inspiring fear, so the martens could play a partial role in reducing their impact. Between 2019 and 2021 a number of organisations led by Gloucestershire Wildlife Trust worked together to reintroduce thirty-five pine martens to the Forest of Dean, some of which have now successfully bred, and further introductions are underway.

*

Boar and pine marten aren't the only lost species to have successfully recolonised British woods. Beavers, too, have mysteriously developed self-sustaining feral populations in a number of Britain's watersheds despite their release only being officially permitted, until February 2025, into enclosures – a restriction that seemed redundant given the success of the big groups living wild, and which has thankfully now been lifted. There are thought to be around 500 beavers on the loose on a number of rivers in the south and south-west of England, and more than 1,000 in Scotland. The positive impact of beavers in the landscape is widely understood, with the efforts of proactive conservationists like Derek Gow having raised the profile of these charismatic rodents. The extensive wetlands created by their dams form ecosystems that seem to bubble over with life, swarming with invertebrates and birds and everything else that responds to the invigorating combination of water, trees, scrub and decaying wood, all spiced by continual change.

I've been lucky enough to visit a number of beaver reintroduction projects in my advisory role and to witness their amazing ability to create wet, woody habitat at scale, their work acting as a shortcut to the wild with astonishing speed. It's never less than exhilarating to stand, always at least ankle-deep in water, among their chaotic tangle of gnawed-over trees, dams of piled wood and resurgent, natural coppice-growth. The only time I've actually seen a beaver, though, was completely unexpected. Travelling home from an Ancient Tree Forum event in Aviemore, I stopped to camp by a little loch on Tayside. Having done absolutely no research about where to spend the night, I was excited enough to find beaver-gnawed stumps of hazel and alder by the water. With their sharp points they looked like chunky, oversized pencils, and there were paired tooth marks in the woodchips scattered around their base. But I was completely agog when, sitting with my pasta at dusk, the steady wake of a beaver cut across the still loch from the mouth of the river that flowed in on the other side. It swam right beneath me, providing a perfect view of its giant tail powering it through the water, before lumbering up on to the bank only a few metres away. Seemingly without a thought for me, it nibbled the grass and soft rush for twenty minutes before pottering on.

The wider debates around beavers are well rehearsed. Some farmers fear their spreading wetlands will impinge on land that could be producing food, which they might; but this tension exposes the need to think in a more sophisticated way about exactly what we want British land to produce and the impact of our diets on its health, as we'll touch on in the next chapter. There are anglers who assume beaver dams will block the movement of fish along rivers, although monitoring has shown that salmon may actually do better when beavers are around; this should be no surprise when we consider that, until a few hundred years ago, almost every river would have had long stretches defined by their activities, and fish evolved in those conditions.[3] There are legitimate concerns around beavers felling the 'wrong' trees, on to roads or in people's gardens, and there are places where their wetlands could affect human infrastructure. But plenty of countries on the Continent manage these issues, following successful reintroductions that have seen beaver numbers across Europe rise from fewer than 1,000 animals to more than a million, with all their associated benefits not just for nature but also for flood management.[4]

Beavers already live on rivers flowing through ancient woodland in the south-west of England and in Scotland and, where the woods are big enough that their work in the valley bottom affects only a small proportion, their presence should be followed by an overall increase in biodiversity from the new, ever-changing structure created by the beavers and that magical meeting of wood and water. Any rare woodland species that require specific existing conditions will persevere undisturbed throughout the rest of the site.

There are, though, well-founded concerns from woodland

ecologists about the potential impact of beavers in some British woods. Given that our ancient woods now cover less than 3 per cent of Britain, the rarest species they support are typically clinging to survival in tiny, isolated refugia within them. Particularly in the temperate rainforests of the west, beavers could adversely affect endangered communities of lichens and other lower plants if they were to fell and flood entire, small oak or hazel woods; this is already becoming an issue in the super-wet 'hyperoceanic zone' on the west coast of Scotland, where the most significant communities of lower plants persist along with the first legally introduced wild beavers.

This particular issue should only affect a handful of small woods, and elsewhere the ecological impact of beavers even in ancient woods is likely to be positive, but it does illuminate the key challenge around our ancient woods and rewilding: their small size and isolation. The charity Rewilding Britain reckons that it's only worth taking a genuine rewilding-type approach in landscapes bigger than 250 hectares, and recognises that in nature reserves of less than 100 hectares intensive conservation management is typically necessary to ensure good ecological condition.[5] Natural England's Ancient Woodland Inventory now lists more than 52,000 ancient woods; when project officer Neil Ford helped me look at the data in late 2024 (midway through the review that brought even more small woods into scope) we saw that around 37,000 of these sites were less than 5 hectares in size. A high proportion of our special ancient woods, then, are tiny, from an ecological perspective, and isolated in farmed or urban landscapes that over recent decades have become increasingly denuded of trees. Less than 1,000 ancient woods in England are bigger than 50 hectares, and only around 300 are bigger than 100 hectares, which emphasises the importance of places like Dovedale and Dodgson Wood.

Their small size and isolation mean that most woodlands aren't big enough for natural processes to drive the dynamism and structural diversity required to ensure the health of their ecosystems; and many of them will, in reality, simply never form part of a wider, rewilded landscape but will remain surrounded by farmland or towns. In these smaller woods that aren't brought into landscape-scale conservation schemes, it's essential that we intervene to create the complexity that our declining wildlife needs. A well-intentioned but misguided approach, inspired by rewilding but without any attempts to introduce dynamism and structure, could simply replicate the neglect that has led to crashing declines in woodland wildlife since coppicing and other traditional management ended. Tom Burns, the woodman at Knepp, explained that a nuanced approach is even necessary at this flagship site. While some of the estate's ancient woods have been incorporated into the wilding project, he's led the restoration of coppicing in others, over the road and outside its influence; they now supply materials for the estate's regenerative farm and market garden as well as providing great habitat.

Even the most active woodlanders, however, can take important lessons from primeval woods like Białowieża and the British rewilding projects, in the many places where we need to continue – or restore – active management by people; their inspiration can help our activities better mimic nature. The most obvious thing we can do is to increase the size of our existing woods wherever possible, pushing their boundaries out and allowing woody ecosystems to expand into the landscape; an effect similar to the way the ancient woods work within the rewilding projects, while also connecting existing woods to one another. The scrub and saplings that develop on the previously open ground around the woodland edge ensure that a future generation of trees develops, creating the complexity that supports the many woodland species that rely on thorny thickets, light and warmth. The core existing areas of woodland that are home to particular species of conservation concern, like rare lichens and bryophytes, can be left well alone where this is best for them or, if needed, managed specifically for those species – to the detriment of regeneration within the wood if necessary.

A phase of something like this happened in the 1980s, when many upland woods like Dodgson and Dovedale were fenced as a result of Professor Pigott's experiment that proved it was sheep preventing a new generation of saplings from growing. At Dovedale Wood, in particular, it was obvious during our summer of surveys that whoever had planned the line of the fence had purposely pushed it as far back from the woodland edge as they could get away with. Now, thirty or more years on, the space created has become a riot of thorny scrub and maturing ash saplings, softening the hard line between habitats and forming a case study in the way woodland naturally claims – or reclaims – open ground. An exciting next step in Dovedale could be to remove the fences altogether, taking a rewilding-inspired approach by introducing light cattle grazing throughout the entire 500-hectare valley and allowing the woodland, grassland and scrub to interact more naturally, each habitat continuously expanding and contracting with the influence of the grazers, and the whole lot intermingling to create much more than the sum of their parts in ecological terms.

There's also a place for sensitive grazing in some ancient woods in our more typical compartmentalised landscape. Low numbers of cattle can help maintain the particular microclimates required by lichens and other lower plants in temperate rainforests, for instance, keeping the understorey relatively open beneath the canopy – an effect that can also benefit some of the special birds of those woods, like the pied flycatcher and wood warbler. At Dolmelynllyn in Eryri, the National Trust rangers graze fewer than a dozen cattle through 200 hectares of woodland and ffridd – the scrubby wood pasture of the Welsh hills – to create the nuanced conditions needed by rare lichens while also allowing saplings to develop. In Dodgson Wood, meanwhile, as well as the rainforest features, wild limes and omnipresent woodland archaeology – it's a busy wood – there's a small remnant population of touch-me-not balsam in a boggy area at the foot of the hill, a rare woodland flower that's the only food plant of the even-more-rare netted carpet moth. The balsam seed needs gently puddled ground in the shade beneath the canopy to germinate, so here, too, the neighbouring farmer occasionally lets a few cows in to maintain the right conditions.

As Guy Shrubsole explores in *The Lost Rainforests of Britain*, though, getting woodland grazing right is something of a dark art: cattle in numbers that create the more open conditions beneath the trees that help support lichens, flycatchers and other species now might also nibble new saplings, depriving a wood of its future, while a lack of scrub excludes a wider range of endangered woodland birds.[6] The immediate answer lies in what's become known as adaptive management, regularly monitoring 'target species' – be they lower plants, birds, flowers or moths – as well as the regeneration of saplings and the wider ecosystem, and tweaking the level of grazing in response, sometimes even skipping a year or two.

The whole thing becomes easier and even more powerful, though, if we can extend the area of the wood, as we've seen – we might not achieve anything like Białowieża's 3,000 square kilometres, but bigger is always better when it comes to creating space for nature. And, wherever possible, we should restore the connections and interrelationships between our isolated ancient woods and any other surrounding habitats, be they heathland, wetland or the rivers that flow by or through them; they're normally all better, like at the rewilding projects, when sympathetic management brings them together, restoring their scale and multiplying the niches they provide for wildlife.

We can also make our woods wilder even where we're driving change with chainsaws rather than with hooves and browsing mouths. This might mean re-wetting them by blocking ditches, leaving more decaying wood even as we create more open space, or welcoming more of the plants that support more wildlife – like bramble, willow, hawthorn and blackthorn. And it might simply be about paying closer attention to the dynamics of more natural woodland ecosystems and trying to replicate them through the trees we choose to fell and the shape of the four-dimensional structures we leave.

In the woods of the Eastern Moors Project in the Peak District, the rangers left their chainsaws behind and winched trees over to let more light in, leaving a beautiful mess that looks just like it was created by a storm – or a herd of ancient super-herbivores. The infamous 1987 'hurricane' that flattened woods across southeast England also informed modern approaches to the more natural management of ancient woods. With Oliver Rackham and George Peterken's work only just beginning to affect policy, woodland management remained heavily influenced by forestry and the instinct in many of the windblown woods was to tidy up the fallen trees and plant new ones, as any good forester would in a plantation. A few places escaped this attention, like parts of Toys Hill in Kent, and thirty-five years on the difference is stark.

Where fallen trees were cleared and replanted there is, with some predictability, a uniform plantation of broadleaved trees of the same age, with few shrub species and little variety, although the ancient woodland ground flora persists beneath. The areas that weren't cleared, however, far from being 'devastated' or 'demolished' as was reported at the time, are energising tangles of life that have a sense of the wild about them: trees that broke in the storm lie in various stages of decay on the woodland floor, while many of those that failed from the roots remain alive and have developed new crowns from their former side branches. The canopy is varied and intermittent, and bramble and shrub species have been able to colonise the gaps, while the holes created by rootplates being torn up have formed little wetlands where they'd quickly filled with water, as I saw in Dodgson Wood after Storm Arwen. Today, we recognise that native woods are evolutionarily adapted to respond to disturbance like this, and that, although there's justified concern about the increase in storm events that climate change might cause, the impact of high winds can be beneficial to the ecology of neglected ancient woods; we typically no longer tidy up.

As we walked wearily and happily down the track to Białowieża's big wooden gate, Renata told us about an old woodman's greeting, a curt little phrase that translated as something like 'luck of the woods'. It was a reminder that, here, a trip to the forest even today is normally still in search of life's essentials: wood, mushrooms, berries, game. I asked if local people would visit just to enjoy it, and was given short shrift; it was apparently unthinkable that the rural inhabitants of the remote village at the park's entrance would come 'for analyse, for observe', although people from other parts of Poland make tourist pilgrimages to the place. I tried the new greeting on an old ranger creaking past on his bike, a basket of mushrooms balanced improbably on his handlebars. He warily replied, 'Dobje,' the most common hello, while giving Renata a quick look that said, 'Where did you find this lot?'

Białowieża may not be the perfect example of a truly natural wooded ecosystem, deprived as it's been, like most places in Europe, of the influence of wild herbivores. But in the long-matured richness of its old-growth habitat and in the intoxicating sense of wildness from knowing we were sharing the wood with bison, boar, lynx and wolves, it provides a tantalising glimpse of an older, less human landscape. Alongside the resurgent scrub of Knepp and the bison-fuelled chaos of the Blean, with their vital reintroductions of keystone species driving more natural processes, we can begin to imagine a different future for our ancient woods that ensures they're as rich as they can be for their struggling wildlife.

Rewilding is an exciting and invigorating approach that has huge potential to change the British landscape and its ecosystems for the better, including by restoring the powerful impact of unpredictability and allowing microhabitats and species interactions to develop that we don't even know about; where we can dedicate whole landscapes to a more natural approach, ancient woodland will have a crucial role to play as a source of biodiversity that can spill out beyond its current artificial edges and repopulate the surrounding land. But in most of our small and isolated ancient woods it's essential that we persevere with, or reintroduce, sensitive active management by people to ensure they're in good condition to support our declining woodland wildlife, and to deliver all the other things we need of them. Even where woodlanders are driving change, though, we need to act on the lessons of rewilding. All our woods can – and should – be wilder.

CHAPTER TWELVE

The Long Spring

Belonging and the new woodland culture

THE CREAMY WOOD curls away in thin strips to gather in a satisfying little heap at my feet, and the rough spoon I'm working has started to sit comfortably in my hand, with the square edges of the axe-hewn blank smoothed off. Something similar is happening in my mind as I focus on the simple, absorbing task, my endless to-do list and wider worries falling away like so many wood shavings. I've spent plenty of my time at work on some of the rougher coppice-crafts – laying hedges using stakes and binders I cut in the woods, or cleaving chestnut poles to build fencing – but my only experience of what you might call higher-end greenwood-working is in my occasional attempts at spoon-carving.

Although my spoons are rugged and misshapen in comparison to the perfectly proportioned specimens I see at woodfairs or on Instagram – characterful, you might generously say – I find both their making and their use deeply satisfying. Carving a spoon is a delightfully tactile and meditative experience as well as a minor life lesson, requiring that you follow the form of the undried 'blank' of green wood as it dictates your knife-strokes: you must always work with the grain, and any attempt to impose something different typically ends badly. The pleasure in carving also, surely, speaks to our ancient relationship with wood, both as a medium and in use; studies have shown that we're healthier and happier when we have lots of wooden things around us in our homes.[1]

I'm far from alone in having found such joy in my naïve efforts, and there's been a boom in interest in spoon-carving in recent years, with a strong online community sharing their efforts and tips on social media, local spoon clubs forming, and even festivals that attract hundreds of carvers. When I went to a spoon fair, a friend took the mick about my 'old man hobby', but that outdated stereotype couldn't be further from the truth; there were people of both sexes, ranging in age from toddlers – whittling with potato peelers – to the older generation, and everywhere in between. And everyone was welcoming and sociable, always ready to strike up a conversation with a newcomer like me.

The community extends to professional coppice-workers who use the summer to transform their winter's harvest into beautiful and useful products, and the resurgence of these ancient crafts has become an important part of the new coppice economy. At Makers Barn, a craft shop in Petworth, West Sussex, there are bundles of beanpoles and pea-sticks alongside woven willow baskets, and Windsor chairs made in much the same way as the bodgers of the Chilterns turned them for centuries. Walking sticks finished with antler handles nestle alongside 'besoms', traditional brooms of bound birch twigs. There are tables full of treen, the small wooden goods that would have been ubiquitous throughout history: cutlery, serving spoons, mugs and soup bowls.

Every product demands to be handled and smelled, and, while the graceful utility of the skilfully made everyday items appeals most to me, other pieces push the boundaries of the woodturner's craft to form delicate artworks. There are decorative plates and platters, and bowls turned from oak burls that feather out to a rough filigree of paper-thin bark around their rims. The old stone building – with its timber-framed roof – is stacked with coppice products like a medieval storehouse; but in reality this is a twenty-first-century incarnation of coppice culture.

The shop is the brainchild of Rosie Rendell, a coppice-worker who began the business to sell her own wares as well as to support other craftspeople working in wood, along with local leatherworkers and potters. Like most modern coppice-workers who manage to make a living from their trade, Rosie depends on adding value to her crop, and her natural entrepreneurship and knack for marketing are as important as her skill with wood, masterful as that is. Rosie also uses the summer months to convert the stakes and withies she's spent the winter cutting into beautiful, sinuous fences, arches, and even raised flower-beds made using the techniques of the old hurdle-makers – a realisation of the prediction in my 1956 Forestry Commission booklet that the best way to make hazel pay would be by adapting hurdles for garden use.

New communities of coppice-workers, and enterprises like Rosie's shop – along with many of the other projects we've visited – mark the onset of a long and drawn-out spring for Britain's ancient woodland, its wild and human life tentatively returning after its gloomy winter during the Locust Years of the post-war era, and the subsequent dark age. Places like Spring Park, Ashtead Common, Forge Wood and Wilder Blean show the wide range of approaches that committed woodlanders have begun to take to rejuvenate our neglected ancient woods, each one a new story of our place in nature threaded through with the sense of belonging those passionate woodlanders find in their work.

Elsewhere, though, more than 90 per cent of our native woodland remains in poor ecological condition and its wildlife continues to decline. While this situation is a result of changes in society over the past couple of hundred years, with their story to this point being so deeply intertwined with human history, the restoration of more of our woods is eminently possible as more new woodlanders develop modern versions of our ancient connection. The continued resurgence of our lost wood culture is essential if we're to return ancient woods to the thriving ecosystems that they should be, and if they're to achieve their potential in providing all the other things we need from them now: carbon storage, flood mitigation, and the health and wellbeing benefits of time spent among the trees – not to mention wood products that can substitute for carbon-intensive materials.

The innovative woodlanders I've met fill me with hope for this woody future. The nascent culture is excitingly youthful – in spirit at least – as well as progressive and diverse, enabling a wide range of people to rediscover that feeling of belonging in the woods, to the benefit of both human and wild life – and in contrast to some of the exclusionary histories of the British countryside. The experience of people like Rosie, and Jamie Simpson, down the road in Dallington Forest, shows the amazing things it's possible for individual woodlanders to achieve, but also hints towards some of the challenges in achieving the scale of active management our ancient woodland requires. Many more people need the opportunity to repair their recently inherited disconnection from woodland, and nature more broadly, and to re-find their home among the trees. The lessons of our ancient woods should also extend beyond their boundaries, helping us make difficult but crucial decisions about the way we use land more broadly. Their history and ecology can inspire the creation and design of new wooded landscapes, ensuring the wider countryside provides much more for nature while restoring trees and wood to the heart of the human story.

In Forge Wood, a few miles east of Makers Barn, Jamie strings black plastic deer netting around his fresh-cut coupes and tidies piles of logs for Andy and his tractor to pull out later in the summer; he'll wait until the young birds have fledged to build even bigger stacks by the track while the ground is dry and firm, ready for the timber wagon. Whether or not we, like Rosie, are turning our attention to the next stages of life for our harvest of wood, the first hints of spring inspire a period of manic activity for all modern woodlanders as we race to finish our winter work before the other inhabitants of the woods return. In the older hazel surrounding this year's coupes, millions of perfect dangling catkins emerge to form a sherbet haze across the understorey; hidden further back down the twigs, the inconspicuous female flowers wait for just enough pollen to land on their tiny purple blooms as it drifts smokily past. We hope to be finished and able to leave the resurgent wildlife undisturbed for a few months by the time the celandines and wood anemones begin to turn their nodding faces to the sun. There's always something for Jamie and his crew of helpers to do in the summer, though, and, among many other maintenance tasks, Graham's bird surveys continue to show the positive impact of their work. As well as the return of nature, the wood has gone from feeling like a remote place – a nowhere that the residents of the hamlet at the old forge simply drove through – to becoming a thriving, human place in its own right once more.

The meaning and sense of belonging – as well as simple joy – that Jamie's community and the greenwood-workers find in coppice culture is reflected in a wider network of new woodlanders, drawn by the appeal of physical work in service of both nature and a sustainable future for people. The benefits of woodland work are increasingly apparent: the Japanese concept of shinrin-yoku, or forest bathing, for instance, is now widely known and, while some have suggested that it just puts a label on what many people with an interest in nature have long understood, the research that has accompanied it provides fascinating concrete examples of its various positive effects.

Phenolic compounds that trees produce to inhibit their own

pathogens have long been used to help develop human medicine – salicin from willow bark, famously, is synthesised for aspirin – but there's increasing evidence that even regularly inhaling certain compounds by spending time beneath the trees can reduce the chances of people developing illnesses. Another study demonstrated that an intensive three-day forest bathing course improved immune system function in people where it was previously weakened, and that the effects lasted at least a week, and in some cases as long as a month. Spending time in the woods has been proven to reduce blood pressure and cortisol, the stress hormone, and the positive impact on mental health, already trusted instinctively by many people, has been quantified and recommendations developed to maximise its effectiveness.[2] It shouldn't really be a surprise; as we've seen, humans evolved in a wood pasture landscape and for generations our very survival depended on an intimate knowledge of – and relationship with – the wooded ecosystems of which we formed but one small part.

Although there are fewer – if any – studies specifically on the wellbeing of woodland workers, I feel sure that the benefits are multiplied. As I've experienced during my career and navigating my own periodic challenges with mental health, woodlanders are subject not only to the life-enhancing effects of time spent in nature but also the benefits of doing work that is physical and meaningful, both of which have independently been shown to be good for us. Like the decline of traditional woodland management, it's a very recent development in the human story that we don't use our hands to do most of our work, and this loss of our physicality is causing an epidemic of health issues associated with sedentary jobs: heart disease, diabetes and back pain, most obviously, but also poor mental health.

I felt the increasing benefit of time spent working in nature throughout my career as a ranger. But it was at its most powerful – even more so than among the limes of Dodgson Wood – after I was flattened playing football while I was working in the White Peak. I went up for a header and, to everyone's surprise, won it.

My opponent, arriving a second late, instead headed me in the face, fracturing my eye socket and causing a severe concussion that lasted months and had me ticking off a checklist of head injury symptoms as they came and went. One day I might feel nauseous and low, the next euphoric but completely forgetful. My weight plummeted, which I attributed to the amount of energy my body was using simply to heal my brain.

The rollercoaster settled down over weeks into the worst depression I've ever felt, anxiety (a newer one for me) and a combination of amnesia and brain fog that left me in tears after a couple of meetings when I returned to work – much too soon, with hindsight. One of the challenges with a brain injury is that it's invisible, and it proved almost impossible for the doctors to assess accurately how I was feeling or how things were progressing – and, having been knocked senseless, I wasn't in a position to judge or explain either. Staying at home in a dark room, as I'd been advised to do, didn't help, as boredom and loneliness made the anxiety almost unmanageable.

Fortunately, the rangers had a long, overgrown hedge to lay on farmland near Dovedale that winter, and in kneeling at my section in companionable silence with the others down the line I found the most effective salve for my shaken brain. My mind could turn at its leisure while I swung my billhook into the backs of gnarly hawthorns until they folded gently and satisfyingly over with a strip of bright wood exposed. The steady pace, the effort in relatively mindless action and my immersion in the microcosm at the bottom of the hedge, with its varied textures and soothing shades – as well as a sociable robin – seemed to sand the edges off the jagged, physical sensations of anxiety and left me feeling more at peace than I had since the accident.

I had, I realised, observed the calming, sustaining power of outdoor work in other people without having understood quite how potent it could be. As a trainee ranger in Cheshire, I worked for weeks with a group of teenage boys who'd been excluded from school and were threatened with youth detention; the

Prince's Trust put together a last-chance programme of activity for them that included practical conservation work. They were, predictably, a wild bunch, led by an ex-marine who fought them each lunchtime, ten on one, to help them let off steam. While it didn't work for all of them, some of the boys settled quickly into the work, cutting rhododendron or digging cobbled drains into paths, the intensely physical labour absorbing their abundant energy and the rich, novel surroundings delighting them. Lots of people, of course, simply aren't cut out for academic work, and I've seen time and again with young volunteers those who don't get on so well with school gaining confidence by doing practical tasks well and enjoying it.

And the same can be true of adults. When I worked at Dunham Massey the rangers formed a partnership with Mind, the mental health charity, who prescribed volunteering on conservation projects for people who hadn't been able to work in years due to struggles with their mental health. Again, of course the impact wasn't universal, but some of the people who came out with us found a new sense of purpose or calm by discovering the solace and satisfaction of using their hands outdoors.

The downside of being involved with nature conservation is that it also increases our eco-anxiety, as we're more exposed to the realities of the crisis in nature and climate, as I experienced most dramatically in Dovedale when I grieved the impact of ash dieback; Aldo Leopold wrote in the 1940s that 'One of the penalties of an ecological education is that one lives alone in a world of wounds.'[3] But taking action in response – in community with like-minded folk – is, in turn, a powerful remedy. Robin Wall Kimmerer compares the need to restore ecosystems to a dinner party; we've enjoyed the 'feast' by exploiting nature, but now we need to do the dishes. She points out, though, that 'anyone who migrates to the kitchen after a meal knows that's where the laughter happens, the good conversations, the friendships. Doing dishes, like doing restoration, builds relationships.'[4]

*

Sad to say, however, these evolutionary benefits of working in – and for – nature haven't been available to everyone over recent generations – and they still aren't today. The stereotype that the woods and wider countryside can be unwelcoming places for people who haven't grown up there, and particularly to female workers and people of colour, is grounded firmly in truth. Every woman I've worked closely with in the woods has told me they've experienced sexism specifically around the fact that forestry was historically a male profession, and during my practical career I never worked with anyone who wasn't white. The social conservatism in one place even extended to a woodlander I know being derided by neighbours for his long hair.

I still come across plenty of rural residents who are suspicious of any 'incomers' to their area, and I've been on the end of it despite my privileged position as a straight, white, middle-class man. A couple of people local to Forge Wood were hostile to Jamie for years: one referred to him as an 'oik' – solely, I think, for his relative youth and South London accent. At worst, I've met people who persist in the belief that only families who've managed the land for generations have the right to decide its future, be they the owners of big estates or of small upland farms. I've seen the balanced perspectives of colleagues – not to mention my own – dismissed more often than I care to remember, and occasionally with alarming aggression, as we've respectfully tried to represent the interests of nature and wider human society.

If we're to rediscover our fundamental societal relationship with woodland, and with the land more widely, then many more people will need the opportunity to build – or rebuild – their own connection with special places like the woods in this book, and to influence their future for the better. We need new ways to articulate the innate potential we all hold to develop these powerful ties as well as to think and talk about how we can find reconnection, and we must ensure that a renewed relationship with land is available to anyone, regardless of their background.

It's clear to me that in working, relatively briefly but intensely, in the woods that I've known, alongside brilliant colleagues and in service of their non-human inhabitants, I've developed a deep personal relationship with them (and when I return it's as though I never left, like picking up with a close friend when I haven't seen them for a while). It's not my birthright but has been born simply of my showing up and paying attention day after day and month after month, for years, and of investing my effort and care.

I don't believe that I've got any more right to this relationship by having worked in those woods; although, to me, working in them has created something very meaningful. I know people who feel just as powerful a sense of belonging with the woods where they walk their dog every morning, or where they go to watch birds or anything else. Most of us probably feel like we belong – or belonged – to the places we played as children, like Eddie's Wood, to the out-of-the-way places where we first went to drink and smoke with our mates as teenagers, or to the urban streets where we necessarily spend the majority of our lives, regardless of whether we'd rather be out among the trees. In *On Gallows Down*, Nicola Chester describes her revelatory sense of belonging to the countryside of the private estate on which she is a residential tenant – despite her access to the place, and indeed her home, being vulnerable to the whims of the owner and their employees. She explains that 'I didn't need to "come from" this place to be part of it. I only needed to get to know it, to let it under my skin on my own terms, wherever I was from, or "coming from". . . Anyone could make a place their home by engaging with its nature.'[5]

Embracing this idea of *belonging*, which has been promoted by the Right to Roam movement, among others, allows us to find meaning in the ancient traditions of woodland management and coppice-crafts, should we wish to, but not to the exclusion of other activities and other interests, nor to the exclusion of people who haven't been born into woodland work or the British countryside.[6] I would perhaps also add something essential about respecting the processes and limits of the land, as Wendell Berry wrote, for those of us involved with doing anything that affects it: he describes his similar philosophy of being a 'friend' to the land, and I believe that this is something that anyone, from anywhere, could choose to be.[7]

Not only is it right and fair that more people have access to woodland work, but the woods need it too. There currently aren't enough people interested in looking after them, which has led to a severe shortage of woodlanders with the necessary expertise, whether it's in woodland ecology and management planning or felling and forwarding – a skills shortage that extends to the wider forestry industry. Diversity in the backgrounds of the passionate people we need to enter the sector will bring equivalent variety and vibrancy to the woods, as we've seen in Jamie's bold work at Forge Wood and in Rosie's novel business supporting other woodland craftspeople; the new woodlanders bring energy and unpredictability akin to the unexpected ecological interactions wrought by the bison in the Blean.

The inclusive nature of the spoon-carvers and the coppice-craft movement reflects a similar spirit among the most exciting modern woodland managers, where we find enlightened woodlanders running their businesses in new ways with a stronger focus on social justice. The Landworkers' Alliance, for instance, is a workers' union that campaigns for a fair system of food production and wider land use, with a focus on both environmental sustainability and a more fair society; their forestry group counts among its number some of Britain's most progressive woodlanders, including a disproportionate number of female workers compared with the wider sector. Working with some of them and seeing their ideas in action has been one of the most energising aspects of my recent work to develop a stronger community of woodland contractors capable of sensitively helping to look after the National Trust's woods.

The ranks of the Alliance include Leeds Coppice Workers, a cooperative that manages woodland on behalf of the city council and Yorkshire Wildlife Trust, but also delivers forest school activities and runs inclusive volunteering days that are both educational and therapeutic. The fact that the eight members work as a cooperative increases their collective skillset and insulates them from some of the financial vagaries of woodland work. Working Woodlands Cornwall, meanwhile, is a Community Interest Company that similarly cares for some of Cornwall Wildlife Trust's woods with a focus on nature and the circular economy; they also run courses to help women and non-binary people experience woodland work and to find their way into the sector.

Of course, lots of more traditional landowners – and other people who were lucky enough to be born into a generational relationship with the land – look after the countryside, and their woods, very well. But, purely in ecological terms, there are still many ancient woods where a focus on rearing pheasants for shooting means their soils and species are in decline (and that people can't visit them), or where simple neglect – through disinterest, misunderstanding or economic pressure – means they're not delivering for nature or people. The flipside to the encouraging trend to begin the restoration of Plantations on Ancient Woodland Sites on land owned by the government and charities is that barely any restoration has begun in woods in private hands.[8]

Every now and then, I even hear of ancient woods being stripped of their standard trees or planted with conifers despite all the changes of the 1980s. In a climate and nature emergency, this can't be right. And while any debate around the future care of our countryside requires great respect for nuance, not to mention for the people involved, it's impossible to approach the subject without considering how few people actually own land. Only about 35 per cent of Britain's broadleaved woodland is in the care of conservation organisations, the government and local authorities, for instance, with the rest being in private hands;[9]

across all types of land, it's likely that less than 1 per cent of the population own half of England.[10]

In Britain, we suffer from a uniquely severe disconnection from nature.[11] It's a product, in part, of that highly concentrated story of ownership, and also stems from the – closely related – disenfranchisement of ordinary people from the land during the period of Enclosure – when the rights of local people to take things they needed from common land were removed, as we saw among Ashtead's oak pollards. The termination of common rights across much of the country was a fundamental shift in the relationship between British society and the fields, woods and wood pasture on which many people directly relied, and it happened earlier and to a greater degree than in most other countries, alongside our extreme experience of industrialisation.

The crisis in nature and climate is also closely integrated with this unjust history, and together they form a depressing spiral of effects that have led us to the state we're in. The abuse of the natural world is a product of an exploitative economy combined with our resulting disconnection from the land; the dominant culture that ensued perpetuates our alienation from nature and allows the unsustainable use of natural resources to continue. And, while we're all affected, the most severe consequences of both economic injustice and nature disconnection continue to be felt by the most marginalised people, both in Britain and abroad.[12]

Where there's a vicious circle, though, there's the potential to reverse things so it becomes a virtuous one. We should be angry about the history of Enclosure, but we can also do something about how land works for everyone now. We need not only a better foundation of grants to support good woodland management by private owners, but also stronger regulation to ensure they yield to the limits of the land, protecting and restoring nature and ensuring the provision of essential ecosystem services for everyone.

Some would argue that the value of our ancient woods for

biodiversity, carbon sequestration and water management is so significant that we should also look to new approaches and attitudes to their ownership – even new forms of ownership – by which they can be better managed for the benefit of all. Perhaps we need to restore some common rights; they might not look like the rights of pannage or estovers – although there could be a fresh role for them, in some places – but should reflect the carbon stored in the woods or their value for nature, so those vital but less tangible assets become a modern common treasury. The last couple of hundred years could become a short blip in our history of the sustainable, communal use of land for 'time out of mind'.

And ultimately we need to build a fairer society where more people – including those people most affected by historic injustice and the climate crisis – can reap the tangible benefits of being reconnected to the land and play their part in restoring our ecosystems as a core component of a sustainable future. Oliver Rackham reminded us that ancient woods are typically older than the parish church, and they would have been just as central to daily life. We need to find ways to restore them to the heart of all our communities.

There are rays of hope in the hundreds of community groups involved with managing their local woods across Britain; from more formal versions of our small-scale approach at Spring Park, where volunteers take firewood in part-payment for their labour, through to big forestry enterprises run collectively, as we find in some parts of Scotland, where community groups now have the preferential right to buy land where doing so will support sustainable development. The Scottish projects that typically make the headlines are the ones where people have come together to buy huge tracts of land for rewilding: the Tarras Valley Nature Reserve, near Langholm, is perhaps the most high-profile example. But I'm just as excited by the more workaday projects that quietly seek to use these new and more equitable models to care for existing woods. The North West Mull Community

Woodland Company has brought hundreds of hectares of neglected plantation into sustainable management, provides all the island's woodchip for biomass boilers, and produces firewood, among a range of other community-focussed initiatives.

Despite this positive step north of the border, less than 1 per cent of British woodland is currently in community ownership, according to Robin Walter's review of the state of our woods in *Living With Trees*, compared to around 20 per cent in both France and Germany.[13] I suspect that it is another symptom of the unique disconnection of British people from nature that the idea of community ownership of woods and land seems almost unthinkable to many people, even as legislation makes it increasingly possible.

If we're to restore all our ancient woodland, projects like that on Mull also show the way in their forestry-inspired ability to achieve work at scale. The new woodland culture won't be able to rely on coppice restoration and greenwood crafts alone. Carved spoons, beautifully turned bowls and hurdle fences keep ancient traditions alive, fostering that sense of belonging and providing vital income for coppice-workers, and there will always be a place for them. But - sadly - the time needed to make them means they're luxuries and won't return to being the everyday essentials they once were, and the demand isn't strong enough to drive the restoration of all our woods.

Selective felling and continuous cover forestry will play a major role in restoring active management at a larger scale, as we've seen, and in making our woods relevant to the everyday lives of more people; estates like Rushmore and some of the woods managed by foresters in the Landworkers' Alliance demonstrate how ancient woods can be brought into sensitive, productive management that more readily pays for itself through new markets for larger-diameter wood. Once again, the best woodlanders bring a number of different skills together, often milling the wood they cut into fencing or planks or even constructing timber-framed buildings – and, in the case of one of member of the Alliance, spotting a gap in the market for flat-packed compost toilet kits.[14]

With the current predominant market for firewood inevitably in decline, I feel sure that the best hope for the sustainable economic management of our ancient woods lies instead in the development of new technologies to bring so-called 'wonky wood' into use for construction and other materials. As we saw beneath Spring Park's oak standards, innovations in processing technology and computer design mean that wood that isn't straight enough to meet the current demands of the timber industry should play an increasing role in future, in laminated beams or through novel uses for knee timber, creating a market for trees cut during coppice restoration and selective felling; we need to rediscover the foundational human tradition of building with the products of our ancient woods, but using inventive new techniques. The use of timber in buildings will connect many more people to the good management of woodland and its ability to provide for our essential needs; these sophisticated innovations could rekindle a feeling of belonging to the woods in people who wouldn't dream of carving a spoon or perhaps even spending time among the trees.

That said, if our ancient woods are to truly fulfil their potential in helping us mitigate the crisis we're in, I feel strongly that we can't leave it to the market alone, essential as wood products will be in a new green economy. Ancient woodland is so special, and so vital for all the other things it provides, that government grants should be enhanced to ensure they cover the costs of its exemplary care, reconnecting everyone to a better future for our woods and their wildlife.

Creating a brighter future for our ancient woods and for people will also mean looking beyond their boundaries, where we need to make difficult decisions about how we best use land to ensure it provides the things we all need. Even if a vibrant new woodland culture was to get all ancient woods into good condition

– which should be a conservation priority – they only cover 2.5 per cent of Britain; we urgently need to restore nature across the rest of our landscapes too, including replacing the trees and woods lost from farms as a result of the drive for increased productivity in the second half of the twentieth century. While we can't create new ancient woodland, projects such as Knepp have demonstrated that it is possible to create the habitat conditions for some endangered woodland species very quickly, like the thickets of scrub that support a resurgent population of birds. We can use the example of ancient woodland to help balance our future use of land so that it provides everything we need from it, while also ensuring that the many new woods being created form the richest possible habitat and contribute to our restored relationship with the natural world.

There's more interest in the creation of new native woodland today than at any time in history, driven by the urgent need to 'draw down' and lock up atmospheric carbon. Creating woodland is widely accepted to be the most effective way to do this, although it's no panacea: our most urgent task is to radically reduce carbon emissions. The British government aims to create 30,000 hectares of new woodland each year to increase tree cover to 16.5 per cent of land area by 2050, as part of a package of measures to achieve net zero carbon emissions.[15] At the National Trust we want to establish 20 million trees by 2030, equivalent to the 18,000 hectares that would increase our woodland area from 10 per cent of our holding to 17 per cent, but spread across a wider range of woodland types including wood pasture and orchards.

Creating new woodland, though, means converting land from something else – typically farmland, as the main use of land – and it's never easy. When I moved to the Lakes I was excited to get out and meet the farmers in my patch, hopeful that we could collaborate to get more nature back onto the historic farms that form such an important part of the landscape and local culture. Some of them welcomed me in, showing me restored hay meadows and trout in gravelly becks, but I also met with resistance. One of them – now retired – memorably told me, stony-faced, that he hoped I'd prove 'less of an arsehole than that other one'. A previous ranger had apparently told the farmer that he wasn't too interested in sheep but loved trees. 'Well I told him,' the farmer said, 'I love sheep and I hate trees, so don't get any ideas about planting any more of the things on my farm.'

With its tight-knit shepherding culture, built around the distinctive Herdwick sheep, and its landscape beloved of so many, the Lake District is a particularly contentious place in which to think about change, and relationships between the farming community and conservationists can sometimes be strained. But I believed then, as I believe now, that by working together we can achieve the things we both want more effectively than we would alone: more wildlife thriving in the fells and fields, an enduring agricultural community and, as a foundation, healthier and more resilient farm businesses. The rich and varied history of the Lake District's small farms remains so alive that the transition back to a more holistic tradition of land management, where trees are grown and valued alongside livestock, can, on my more optimistic days, seem tantalisingly close – and things have progressed a long way since that conversation more than ten years ago.

This tangible sense of history serves as a reminder that the typical current focus of any one farm on meat, or dairy, or growing crops, is a recent development. Although our landscape has been gradually compartmentalised and simplified for centuries, the most radical changes came in response to the shortages of the Second World War, when government policies understandably drove farmers to focus on maximising food production and the deployment of new technologies like bigger tractors and the application of pesticides and fertilisers. Before that, most farms were 'mixed' farms, growing a range of livestock and some vegetable or arable crops; in the Lakes, the pollards still found in the fields of the valley bottoms and in rougher areas of wood pasture were also important. The poles were cut in summer and dried to contribute to an essential supply of winter fodder in these 'closed' systems, capable of supporting only as many animals as the land could sustain without importing feed – another development of the post-war era.

The farm economy often extended beyond what we'd think of as agriculture to include other rural enterprises, with many farmers in the Lakes also being involved with quarrying or mining, and most farms featuring a 'spinning gallery' where the Herdwicks' thick fleece was converted to wool. Many of the area's ancient woods, too, were integral to the way farmers lived, with coppicing providing firewood, hurdles and parts for tools and equipment, as well as a lost coppice product: rick pegs, scraps of hazel that held hay ricks in place when the grass was stacked to dry in late summer. Stakes were needed in huge quantities to maintain the hedges that created a much denser landscape of small fields, before they were ripped out to enable bigger machinery on to the land. The distinction between farmers and foresters stems only from the point at which these traditions collapsed, and attitudes like that of the frosty farmer who told me he hated trees are a product of just the last few decades.

Many of the farmers I worked with during my time in the Lakes, though not all, seemed to share his sentiment to some degree, following decades of governmental encouragement and financial support to focus on increasing the productivity of their sheep. I came to get on alright with that farmer, despite our sobering initial conversation. I really admired them all and loved spending time in their fields and barns, learning about their work and, as trust developed, figuring out ways we could help one another. They were passionate and incredibly hard-working, and had a quiet but unmistakeable connection to the land; you could never deny that most of them 'belonged' to the landscapes they worked. But, through no fault of individuals, it was sometimes possible to observe that the natural processes of the land had been eroded or broken, and that we, as a society, have asked the farmers to push places like this beyond their limits.

The hangover of the post-war drive to maximise food production was exacerbated by perverse outcomes of the Common Agricultural Policy when Britain joined the European Economic Community. Between the 1970s and the 1990s 'headage payments' encouraged the expansion of grazing to levels the land couldn't support by paying a subsidy for each animal. For now, the Basic Payment Scheme in England continues to pay farmers a flat rate for every acre and props up systems that aren't economically sustainable, let alone environmentally sound. Lots of the farmers I knew were clear in their conversations with me that they saw their role as being to grow as much food as possible, and, although most of them also signed up for additional grants to restore habitats, some of them did so with reluctance or a perceptible feeling of bemusement. A big part of my job was helping to plan and deliver these commitments, typically things like restoring neglected hedgerows, fencing out streams and planting scattered trees; some farmers were keen and took pride in the work we did together, but others admitted that their hearts weren't in the projects and were glad to leave me to it.

Sometimes things got really heavy. I saw the extreme pressures that farmers are under today and worried for a couple of the men I grew to know well as they shared their anxieties and frustrations. One came close to tears explaining the weight he carried as he simply tried to do the best he could under competing influences and demands. He felt a painful shift in public sentiment away from farmers, while his own fields were seemingly overrun with tourists. Environmentalists – 'people like you' – would turn up every week wanting him to try something different. The politics of neighbourly relations were fraught as each tried to navigate their way. And all this underpinned by constant stress as he balanced the high cost of imported feed and other essentials with the low price he received for his sheep. It was heartbreakingly clear that he couldn't carry on doing things the way he'd learned from his parents, but that he found it hard to imagine an alternative.

More than a decade on, change is happening with increasing pace, and there's a chance that it will make for happier farmers as well as a healthier landscape. One of my most encouraging visits in recent years was to a pair of longstanding Lake District farmers who have restored the natural course of the streams that cross their land as they flow into Ullswater, sown wildflower meadows, and planted thousands of trees to create wood pasture and woodland, while still enjoying their – smaller – flock of sheep. They told me that their decisions had been driven as much by financial reality as sentiment, and described the sobering realisation that all the food they produced each year, in the shape of a few dozen lambs, would fit in their farmhouse kitchen. But they've also grown to love the new life on their farm and take satisfaction from the return of barn owls and songbirds to their scrubby edges and new woods, which has fostered a deeper and richer connection with their land.

And this approach is increasingly common. James Rebanks, the influential farmer and writer, does something similar in the next valley; his first step into a more nature-friendly farming system came with the well-funded offer to plant new woods along his streams to slow the flow of water, vital in this landscape following catastrophic flooding in Glenridding and Cockermouth. Having initially been attracted by the amount of essential fencing the scheme would pay for, he too came to think, by the time he wrote *English Pastoral*, that 'there is something about planting trees that feels good. If you have done it well, it will outlast you and leave the world a little richer and more beautiful because of your efforts.'[16]

With the Basic Payment Scheme due to wind down to nothing by 2027, there's typically now an economic imperative for farmers on marginal land like this to diversify away from sheep – without subsidies often a loss-making enterprise – and to focus more on those payments for restoring habitats and providing other ecosystems services like flood mitigation. There's lots of evidence to show that this is the best use of our least productive land for society as a whole: the National Food Strategy, for instance, shows that the 20 per cent of Britain that is least well suited to growing food produces only 3 per cent of the calories we eat, and it would be doing more for us as a nation if it were much richer in nature, stored more carbon and held more water.[17] But Rebanks also notes that 'none of this is at odds with breeding a great flock of sheep or a herd of cattle', the stock-rearing that's core to the identity of many Lake District farmers, while he's also found that it simply makes his life more enjoyable to work with the grain of nature and to see wildlife return.

Although I recognise the challenges, I feel sure that a better future is possible in the Lakes where many more farms look like this, with trees and woods an intrinsic component once again and even more farmers valuing their habitats in the same way they currently do their sheep – while maintaining important local traditions and the character of the landscape. Plenty of other farmers are already beginning to do something similar, or at least engaging with the possibilities. Wood pasture is not only good for wildlife but can provide shelter and shade for livestock; studies show measurable improvements in animal health. I once stood with Sam, a younger farmer I knew in the Lakes, as he gazed lovingly at his Belted Galloway cattle and unconsciously reflected what I'd observed at Ashtead: 'They look well among the trees, don't they?' He joins many other farmers working National Trust land who are beginning to realise the benefits of creating wood pasture, with thousands of hectares having been established since 2020.

Tree hay could once again become an important source of winter fodder as farmers look to reduce their reliance on expensive imported feed. And new farm woodlands can provide fence-posts, firewood and stakes for the many replanted hedgerows and a renewed interest in hedge-laying, not to mention bringing grant income in return for the wider benefits they provide to us all. As well as re-finding uses for those traditional woodland staples, we can also use the lessons of ancient woodland to ensure our new woods provide the richest possible habitat, whether by designing in dense thickets of scrub, enriching glades and rides with wildflowers, creating wetland features, or simply by supporting natural colonisation to do the job on our behalf.

While it's perhaps easier to conjure this bucolic vision in the Lakes, with its thin places in time making the billhook-wielding lives of previous generations seem close enough to touch, a similar future is possible in the lowlands too. In some places, the reintegration of trees on farms might look more formal, with rows of 'agroforestry' trees planted in arable fields to produce a second harvest while also helping retain water and improve the soil, and new – nature-rich – plantations providing timber for future generations. The thriving interest in regenerative agriculture, with its focus on ecological sustainability and soil health, is already seeing lots of trees planted on progressive farms across the UK. Their roots and decomposing leaves are helping to kick-start the slow process that soil expert Felicity Roos described to me of restoring depleted farmland soils to something more like the rich ecosystems found beneath ancient woodland. But there's also scope to create lots more woods simply as homes for wildlife, and for society to pay farmers to expand and connect their ancient woods for those other less tangible benefits like carbon and water management, services that are no less vital to our future than food production.

Modelling by the Climate Change Committee, the government's independent advisers, showed that the space for much more woodland can be found by reducing our consumption of meat and dairy, both of which take up a massively disproportionate amount of land for the calories they contain, as well as by cutting food waste and making improvements in agriculture; under their model, enough land could readily be made available to facilitate the government's commitment of increasing woodland cover to 16.5 per cent of our land area while growing more food for the nation and locking up millions of tons of carbon dioxide.[18]

Essential changes to the way we use land, though, won't happen through evidence and policy, or even because of what they look like on a balance sheet; we'll need to find ways to ensure trees and woodland work for farmers and their interests, as James Rebanks and others have discovered. We need a new – or perhaps renewed – idea of what it means to be a farmer, one that fosters an even richer connection with the land and restores farmers' capability and confidence to tend trees and other habitats in the same way they do their livestock and crops – and with the same pride and passion.

Although many woodlanders use the summer to focus on adding value to their winter's harvest and maintaining the delicate balance of their various streams of income, for all of us the woods are home throughout the year; their very familiarity is in their constant change through the seasons and over years and beyond. While I have been left elated by occasional, memorable wildlife spectacles, like the purple hairstreaks I found myself among in the top of a Spring Park oak, or the flock of willow warblers soundtracking the glistening haze of floating seed at Ashtead, my overwhelming feeling towards the woods I've come to know well – and of British woodland more generally – is less like the giddy rush of love at first sight and more like a down-to-earth, familial affection. The majority of humans in Britain were directly dependent on this homely and unsentimental everyday relationship with woodland, wood pasture and other trees for thousands of years, from the Bronze Age until just a couple of generations ago.

As Patrick Joyce describes in *Remembering Peasants*, his moving history of our overwhelmingly rural, self-sufficient forebears, nature was, for those countless previous generations, neither a garden to be tamed and exploited nor a wilderness to be feared or to stand in awe of, the two perspectives by which we typically view it today thanks to our recent disconnection. Our ancestors were not in nature, but of it; an outlook that we would do well to rediscover, albeit in new and forward-thinking ways.[19] Cultural historian Joe Moran pithily summarises Wendell Berry's similar philosophy: 'The purely natural and the purely human are to him equally undesirable states. There is nothing wrong with using nature for our own ends, so long as we see that we also have a duty to live within its limits.'[20]

In re-finding our place in nature, it's essential that we don't just look back. If there's one thing my experience has taught me, it's that there's no one right way to care for every wood, or even any single wood: the restoration of coppice management will certainly be the best thing we can do in some places, but in others selective felling or the incorporation of ancient woods into big, wilder landscapes will be more effective ways to rejuvenate woodland wildlife and ensure the health of their ecosystems. We'll need to embrace innovation and new technology as well as recovering methods that have served us well in the past, finding 'old ways to achieve new goals', as Sue Clifford, of the charity Common Ground, writes – after noting that 'working woods are edging back into fashion'.[21] Our woods can never return to how they were at any point in the past; by restoring our ancient relationship with them, our job is to make them as healthy and rich as they can be now and for the future.

More people – be they farmers or city dwellers, or anyone else – can rediscover our evolutionary sense of belonging in Britain's ancient woods as we bring them back into good management and take their lessons into the wider countryside. As well as helping to mitigate the existential threat we face from the crisis in nature and climate, a new woodland culture will enable anyone – hopefully everyone – to reconnect with a daily experience of living and working among trees, forging new connections with nature and with one another. Instead of simply being places to visit, the woods that were so central to our story until just a few decades ago will become home once more.

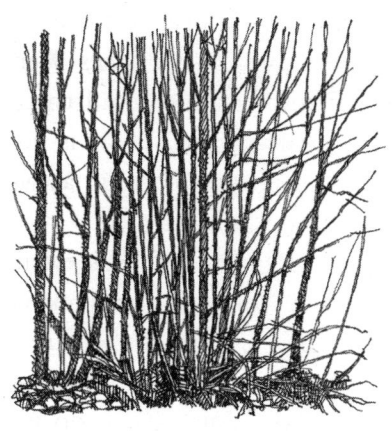

Epilogue

I SENSED, BEFORE I saw, that something was different, as I beat my usual route from the hidden parking spot by Coniston Water, up through the ferns along the edge of the steep ghyll and towards the giant, sprawling small-leaved lime above the charcoal hearth. Even before it came properly into view things felt awry, disorientating in ways I couldn't at first place: I wondered for a moment if I'd thoughtlessly followed the wrong beck. There was too much sky, and somehow too much green filling the space beneath the oaks. Suddenly, surreally, it made sense. The fattest stem of the lime, the one that arched out like a cantilevered sun-shade to spread its canopy above the hearth, had torn away from the stool on the crag and its giant crown now rested on the platform, covering it entirely and even bulging over the sides, slumped under its own weight.

The failure was fresh: despite the fall having pulled the tree's roots clear of the shallow earth the perfect little heart-shaped leaves were all plump and full, not wilted and dry as they would inevitably become, and the scar at its base revealed shining rock and black soil. I could almost hear the lime ticking and crunching as it continued to settle.

Fighting my way in among the crown, the night's rain soaking me where it was held on the leaves, I wondered at the architecture and detail of branches I'd only previously glimpsed by peering 20 metres up. I clambered onto the big, grey-black trunk and balanced along where it bridged the beck to stand above the waterfall. Even the familiar features of the ancient organism felt redefined in the changed light of the open sky. The big broken pole, folded halfway up and twisted open like a rope before my time, now seemed to form the centrepiece of an entirely new, unfamiliar place, accompanied only by one last remaining standing trunk, its smaller crown leaning apologetically the other way.

There's always a thrilling energy when you find a fresh fallen tree; it's even more powerful when it's a tree you've come to know well, the sensation enriched and complicated by surprise and sadness into conflicting layers of emotion. And I was sad to see this one go. It had become a touchstone of my regular trips to Dodgson Wood, a tree I would visit in passing even if I was exploring other ghylls, long after my photocopied map had disintegrated, and a place I'd bring my friends; even those of them who weren't interested in trees would be wowed by the way its arcing canopy framed the platform and the lively water. The timing of its failure felt weirdly portentous: this was my last visit before leaving the Lakes for the White Peak and Dovedale. But I've been back to Dodgson at least every year since, and, having venerated the big lime in the form I originally knew it, I've been amazed to witness its theoretical deep-time regeneration unfolding in front of my eyes.

Within a couple of years of that biggest, most shocking failure, the last remaining standing section broke away, too, springing clear as it fell so that I found its crown nestled in the ghyll, having bounced and slid down the slope. The entire tree, really four

EPILOGUE

or five big trees and more smaller stems, was splayed out from the broken stool at its heart beneath a dome of sky, an area half the size of a football pitch opened among the surrounding oak. The difference between the cavernous, enclosed place I had first come to know, with the lime's solid pillars of trunk and roof of delicate, geometric twigs, was so great as to be almost incomprehensible. The sense of exposure, the envelopment of the charcoal hearth and the chaos of decaying trunks and quickly crumbling dead crowns erased any sense of human scale.

It remained captivating, nonetheless. Another year or two on, the flood of light had drawn forth hundreds of new shoots from the broken base of the stool, the tree's evolutionary ability to coppice having produced new growth taller than me that hid the old stumps behind an exuberant thicket. The irrepressible energy of lime, relentlessly driving its survival and renewal, was palpable. As well as the spring from the stool, some of the lime's broken crowns whose roots retained a connection to the soil now formed lush, low layers of foliage, their leaves reoriented to face the sun. The open spaces between the fallen stems were quickly colonised by bramble and bilberry and ferns, greedy for the light and warmth.

Every year, as all lime nerds must, I looked for evidence of surviving fallen branches having sent roots down into the rocky earth to foment new trees that would see the lime begin to walk through the landscape again. Each time, for years, I was left disappointed, until finally in the autumn of 2023 I scrambled down beneath the big stem that still bridges the ghyll. The stream was as dry as I've ever seen it, a thin trickle that I could splash through in my boots. A handful of branches from the crown of the folded, unwound lime-stem had been driven into the earth and now, with the water low, it was possible to see that banks of sand and pebbles had been deposited behind them, burying them solidly. Their tips had bent skywards and made for the light, presumably forced harder by the depth of the ghyll's steep walls, and now formed a little grove of young trees. Looking closely, I saw that they were thicker than the original branches curving down, a sure sign that they had taken root and will grow to form thriving new individuals if – or when – the connection is broken with their ancient parent. The big lime has taken its next step.

Experiencing the tree's collapse and the beginning of its resurgence brought the ecology of lime even more vividly to life for me. I came to realise that I typically only observe a snapshot in time when I hunt in the woods for signs of hidden limes reinventing themselves, a mere hint of the constant cycle of reiteration that it's rare, on a human timescale, to witness playing out. Stood in the beck that day, I was struck by the thought that I had been observing, over a brief decade or so, a process that has been happening in Dodgson Wood – and in the other remnant populations of native lime – since trees recolonised these landscapes after the last Ice Age. It was possible for a moment to visualise how this tree, and the dozens of other trees along these steep streams, had done something similar over the last few thousand years, an endless churn of failure and decay and new life in all its forms. I lost myself in the realisation that down here, away from the charcoal hearths and oak, my surroundings in the realm of wild lime might have looked much like this at almost any point in that history, or at least that these ghylls have always been defined by variations on this eternal theme of transformation and change.

It extends the other way, too. Wild limes like these will probably outlast us all, and could continue to thrive here in almost any of the futures open to humanity: from hopeful versions of the story, where we figure out our fractured relationship with the planet and thrive for generations more, to all but the bleakest outcomes that might arrive sooner. In a strange way that isn't always comfortable, my tentative understanding of the limes' history and their probable future reassures me when eco-anxiety and climate dread get me down: nature will persist, whatever happens to people. The ecosystems we know now might change beyond recognition, but some species will survive – perhaps including the lime – and the evolutionary branches of the tree of life will diverge and spread again to repopulate the land with new and different organisms, further beyond our imagination than the dinosaurs.

I find myself quickly countering that thought, though. Regardless of nature's ability to recover, there's something deeply shameful about our responsibility for the sixth mass extinction; it is arrogant and staggeringly cruel, careless in the worst possible sense of the word. Gary Snyder, the American ecologist and poet who inspired me during my studies in California, once told an audience that nature 'will remain long after our demise' and 'doesn't need us to save her'. When challenged, 'then why work to stop the destruction?', Snyder – according to fellow writer Jim Dodge – 'grinned hugely ... and replied without a quiver of hesitation, "Because it is a matter of character." Then, with an absolutely wild glitter of light in his eyes, added, "And it's a matter of style."'[1]

We are only one species among millions on our unique and extraordinary living planet and, as easily the most influential, it's about time we showed some style.

The crisis in nature and climate has also, of course, already affected millions of people, and its impacts for humans will only get worse. Anybody with a shred of compassion, not to mention any concern for our own children and grandchildren, must see nature recovery and the mitigation of climate change as the biggest challenge of our lifetimes. Our ancient woods have been in a unique form of stasis for the last few decades as our disconnection from them has caused crashing declines in their wildlife; their story is central to Britain's uniquely severe nature crisis. Like spotting the signs of lime regenerating without truly understanding the constant process of change, we've typically only seen a snapshot of our woods, frozen in time. We've become accustomed to their cool shade and silence, and assumed, because it's all we've known, that this is the way they've always been.

In the same way that I came, from my initial shock and sadness, to see the hope and vitality in the collapse of the big lime at Dodgson, we need to embrace radical change in our ancient woods as we restore their complexity and reintroduce light, warmth and the resultant abundance of life: from rare wildflowers locked in the seed bank and clouds of insects drawn to willowherb and meadowsweet, to thickets of scrub that support struggling woodland birds – as well as new generations of trees that ensure the very future of the woods themselves.

The new wood culture that will support this essential renaissance isn't simply a nostalgic recreation of the past but a vibrant modern version with a life of its own, albeit one that's rooted in the traditions that came before and have persisted from prehistory. Our ancient woods should become unsentimental places of bramble scratches and sore shoulders for a new and expanded generation of woodlanders, but also places of beavers and boar, and where the life-sustaining crash of falling trees comes from the sheer scale of wilder woods as well as from a revival of woodland work.

Ancient woodland is not the wilderness we might sometimes imagine it to be, but neither is it a resource to be tamed and managed just for human benefit; its story lies somewhere between the two, braided irrevocably with our own. In recognising that we are all woodlanders we can rediscover our ancient sense of belonging among the trees, restoring our essential role in the life of the woods as well as their role in ours. And even as we work to rejuvenate the vital remnants of ancient woodland, the lessons we take from its history will inform a healthier approach to land everywhere as we relearn the need to respect its limits and support its processes, recalling that we are of nature, not just in it, and repairing our fundamental relationship with the natural world.

Further Reading

Keith Kirby, *Woodland Flowers: Colourful Past, Uncertain Future (British Wildlife Collection)* (London: Bloomsbury, 2020).

Peter Marren, *Woodland Heritage* (Exeter: David & Charles, 1990).

George Peterken, *Trees and Woodlands (British Wildlife Collection)* (London: Bloomsbury, 2023).

Oliver Rackham, *Trees and Woodland in the British Landscape* (reissued, London: W & N, 2020).

Oliver Rackham, *Woodlands* (London: Collins, 2006).

Ian Rotherham, *Ancient Woodland: History, Industry and Crafts* (London: Shire Library, 2013).

Soil Association, *Regenerative Forestry: Forestry and Forests for the Future* (Soil Association, 2022).

Guy Standing, *Plunder of the Commons: A Manifesto for Sharing Public Wealth* (London: Pelican, 2019).

Tori Tsui, *It's Not Just You: How to Navigate Eco-Anxiety and the Climate Crisis* (London: Gallery UK, 2023).

Charles Watkins, *Woodland Management and Conservation* (Exeter: David & Charles, 1990).

Charles Watkins and Keith Kirby, eds, *Europe's Changing Woods and Forests* (Wallingford: CABI, 2015).

Woodland Trust, *Ancient Woodland Restoration: An Introductory Guide to the Principles of Restoration Management* (Woodland Trust, 2020).

Notes

Introduction

1 Department for Environment, Food and Rural Affairs, *Keepers of Time: ancient and native woodland and trees policy in England* (DEFRA, 2022).
2 Rob Fuller and Martin Warren, *Coppiced Woodlands: Their Management for Wildlife* (Nature Conservancy Council, 1990).
   George Peterken, *Woodland Conservation and Management* (London: Chapman and Hall, 1981).
   Peter Buckley and Jenny Mills, 'Coppice Silviculture: From the Mesolithic to the 21st Century', Charles Watkins and Keith Kirby, eds, *Europe's Changing Woods and Forests* (Wallingford: CABI, 2015).
3 Forestry Commission, *National Forest Inventory: Woodland Ecological Condition – Executive Summary* (Forestry Commission, 2020).

Chapter One: The Pioneers

1 Oliver Rackham, *Woodlands* (London: Collins, 2006).
2 Keith Kirby, *Woodland Flowers: Colourful Past, Uncertain Future (British Wildlife Collection)* (London: Bloomsbury, 2020).
3 Peter Marren, *Woodland Heritage* (Exeter: David & Charles, 1990).
4 Derek Niemann, *A Tale of Trees: How Britain Nearly Lost its Ancient Woodland* (London: Short Books Ltd, 2016).
5 https://www.woodlandtrust.org.uk/protecting-trees-and-woods/threats-to-woods-and-trees
6 C. Reid et al., *State of the UK's Woods and Trees 2021* (Woodland Trust, 2021).

NOTES

Chapter Two: Turn, Turn and Turn Again

1 George Peterken, *Woodland Conservation and Management* (London: Chapman and Hall, 1981).
2 Oliver Rackham, *Trees and Woodland in the British Landscape* (reissued, London: W & N, 2020).
3 Roland Ennos, *The Wood Age* (London: William Collins, 2021).
  Peter Marren, *Woodland Heritage* (Exeter: David & Charles, 1990).
4 Peter Buckley and Jenny Mills, 'Coppice Silviculture: From the Mesolithic to the 21st Century', Charles Watkins and Keith Kirby, eds, *Europe's Changing Woods and Forests* (Wallingford: CABI, 2015).
5 Roland Ennos, *The Wood Age* (London: William Collins, 2021).
6 Oliver Rackham, *Trees and Woodland in the British Landscape* (reissued, London: W & N, 2020).
7 Oliver Rackham, *Woodlands* (London: Collins, 2006).
8 James Arnold, *The Shell Book of Country Crafts* (London: John Baker, 1968).

Chapter Three: Being Cruel to Be Kind

1 George Peterken, *Woodland Conservation and Management* (London: Chapman and Hall, 1981).
2 Rob Fuller and Martin Warren, *Coppiced Woodlands: Their Management for Wildlife* (Nature Conservancy Council, 1990).
3 Oliver Rackham, *Woodlands* (London: Collins, 2006).
4 Robin Walter, 'Thinning', Adrian Cooper, ed., *Arboreal* (Beaminster: Little Toller Press, 2016).

Chapter Four: Common People

1 Oliver Rackham, *The Last Forest* (London: J. M. Dent and Sons, 1989).
2 Keith Kirby and Charles Watkins, 'The Forest Landscape Before Farming', Charles Watkins and Keith Kirby, eds, *Europe's Changing Woods and Forests* (Wallingford: CABI, 2015).
3 https://foundationforcommonland.org.uk/a-guide-to-common-land-and-commoning
  Guy Standing, *Plunder of the Commons: A Manifesto for Sharing Public Wealth* (London: Pelican, 2019).
4 George Peterken, *Trees and Woodlands (British Wildlife Collection)* (London: Bloomsbury, 2023).

ANCIENT

5 Simon Fairlie, 'A Short History of Enclosure in Britain', *The Land*, Issue 7 (2009).
6 Oliver Rackham, *Woodlands* (London: Collins, 2006).
7 Ancient Tree Forum, *Invertebrates and Ancient Trees*: https://www.ancienttreeforum.org.uk/ancient-trees/ancient-tree-ecology-wildlife/invertebrates
8 Oliver Rackham, *The History of the Countryside* (London: Weidenfeld & Nicolson, reissued 2020).

Chapter Five: A Glimpse of the Wildwood

1 Tiffany Francis-Baker, *The Bridleway: How Horses Shaped the British Landscape* (London: Bloomsbury Wildlife, 2023).
2 https://knepp.co.uk/rewilding/wildlife-successes/nightingales
3 Benedict Macdonald, *Rebirding: Restoring Britain's Wildlife* (London: Pelagic, 2020).
Isabella Tree, *Wilding: The Return of Nature to a British Farm* (London: Picador, 2019).
4 Frans Vera, *Grazing Ecology and Forest History* (Wallingford: CABI, 2000).
5 Sophie Yeo, *Nature's Ghosts: The World We Lost and How to Bring it Back* (Manchester: Harper North, 2024).
6 David George Haskell, *The Songs of Trees: Stories from Nature's Great Connectors* (New York: Viking, 2017).

Chapter Six: Woodbanks and Walking Trees

1 Ian Rotherham, *Ancient Woodland: History, Industry and Crafts* (London: Shire Library, 2013).
2 Peter Marren, *Woodland Heritage* (Exeter: David & Charles, 1990).
3 Oliver Rackham, *Trees and Woodland in the British Landscape* (reissued, London: W & N, 2020).
4 Robin Wall Kimmerer, *Braiding Sweetgrass: Indigenous Wisdom, Scientific Knowledge and the Teachings of Plants* (London: Penguin, 2020).
5 Kathleen Jamie, 'A Lone Enraptured Male', *London Review of Books*, 30:5 (2008).
6 Wendell Berry, 'A Native Hill', *The World-Ending Fire* (London: Allen Lane, 2017).
7 Donald Pigott, *Lime-trees and Basswoods: A Biological Monograph of the Genus Tilia* (Cambridge: Cambridge University Press, 2012).

8 Oxford Archaeology, *East Coniston Woodland, Cumbria: Historic Landscape Survey Report* (Oxford Archaeology, 2010).
9 Mark Bowden, *Furness Iron* (English Heritage, 2000).
10 C. D. Pigott, 'Regeneration of oak-birch woodland following exclusion of sheep', *Journal of Ecology* 71 (1983).
11 Nicola Chester, *On Gallows Down: Place, Protest and Belonging* (London: Chelsea Green, 2022).
12 Donald Pigott, *Lime-trees and Basswoods: A Biological Monograph of the Genus Tilia* (Cambridge: Cambridge University Press, 2012).

Chapter Seven: The New Dark Age

1 Accredited Official Statistics: Wild Bird Populations in the UK and England, 1970 to 2023, DEFRA https://www.gov.uk/government/statistics/wild-bird-populations-in-the-uk/wild-bird-populations-in-the-uk-and-england-1970-to-2023
2 Rob Fuller and Martin Warren, *Coppiced Woodlands: Their Management for Wildlife* (Nature Conservancy Council, 1990).
George Peterken, *Woodland Conservation and Management* (London: Chapman and Hall, 1981).
3 Sven-Erik Åström, 'Britain's timber imports from the Baltic, 1775–1830', *Scandinavian Economic History Review*, 37:1, 57–71 (1989).
4 Russell Meiggs, *Home Timber Production (1939–1945)* (London: Crosby Lockwood, 1949).
5 https://www.countryfile.com/wildlife/trees-plants/history-of-britains-forests-and-woodlands-celebrating-100-years-of-the-forestry-commission
6 Forestry Commission, *Bulletin 27: Utilisation of Hazel Coppice* (Forestry Commission, 1956).
7 Derek Niemann, *A Tale of Trees: How Britain Nearly Lost its Ancient Woodland* (London: Short Books Ltd, 2016).
8 Forestry Commission, *National Forest Inventory: Woodland Ecological Condition – Executive Summary* (Forestry Commission, 2020).
9 R. Fox et al., *The State of the UK's Butterflies 2022* (Butterfly Conservation, 2023).
10 Jeremy Thomas and Richard Lewington, *The Butterflies of Britain and Ireland* (London: British Wildlife Publishing, 2014).
11 R. Fox et al., *The State of the UK's Butterflies 2022* (Butterfly Conservation, 2023).
12 Keith Kirby, *Woodland and Wildlife* (Epping: Whittet, 1992).

13 Duff Hart-Davis, 'Country Matters: A Tree Loving David Beats the Goliaths', *Independent*, 2 April 1993.
14 Daniel Alder, Rob Fuller and Stuart Marsden, 'Implications of transformation to irregular silviculture for woodland birds: a stand wise comparison in an English broadleaf woodland', *Forest Ecology and Management* 422, 69–78 (2018).

Chapter Eight: Darkness and Light

1 Alistair Hotchkiss, *Assessing air pollution impacts on ancient woodland: ammonia* (Woodland Trust, 2019).
2 Oliver Rackham, *Woodlands* (London: Collins, 2006).
3 J. R. Madden and R. B. Sage, *Ecological Consequences of Gamebird Releasing and Management on Lowland Shoots in England: A Review by Rapid Evidence Assessment for Natural England and the British Association of Shooting and Conservation* (NEER016) (Natural England, 2020).
4 https://cdn.forestresearch.gov.uk/2022/02/presence_of_rhododendron_in_british_woodlands.pdf
5 https://naturescalendar.woodlandtrust.org.uk/analysis/research-reports/published-research/tritrophic-phenological-match-mismatch-in-space-and-time
6 Stephen Moss, 'Climate disruption to UK seasons causes problems for migratory birds', *Guardian*, 16 May 2024.
7 https://naturescalendar.woodlandtrust.org.uk/analysis/research-reports/published-research/uk-plants-flowering-a-month-earlier-due-to-climate-change
8 Gary Snyder, 'Ancient Forests of the Far West', *The Gary Snyder Reader* (Berkeley: Counterpoint, 1999).
9 Alice Broome and Ruth J. Mitchell, *Forest Research Note 29: Ecological impacts of ash dieback and mitigation measures* (Forestry Commission, 2017).
10 Lisa Samson, *Epitaph for the Ash* (London: Fourth Estate, 2018).

Chapter Nine: Inheritance

1 Archie Miles, *The British Oak* (London: Constable, 2013).
2 Oliver Rackham, *Woodlands* (London: Collins, 2006).
3 https://www.friendsofnotredamedeparis.org/cathedral/artifacts/roof
4 https://www.guedelon.fr/en

NOTES

5   George Peterken, *Woodland Conservation and Management* (London: Chapman and Hall, 1981).
6   James Arnold, *The Shell Book of Country Crafts* (London: John Baker, 1968).
7   Susan Bell, 'How did they do it? From woodland to water', *Wood Culture* 2024 (Woodland Heritage, 2024).
    https://albaola.org/en
8   Archie Miles, *The British Oak* (London: Constable and Robinson, 2013).
9   Oliver Rackham, *Woodlands* (London: Collins, 2006).
10 James Arnold, *The Shell Book of Country Crafts* (London: John Baker, 1968).
11 Archie Miles, *The British Oak* (London: Constable, 2013).
12 Forest Research, UK Wood Production and Trade: 2023 Provisional Figures (2024) https://cdn.forestresearch.gov.uk/2024/05/UKWPT-2023-provisional-figures.pdf
13 https://waughthistleton.com/black-white-building
14 Jeremy Thomas and Richard Lewington, *The Butterflies of Britain and Ireland* (London: British Wildlife Publishing, 2014).

Chapter Ten: Treading Softly

1   Peter Marren, *Woodland Heritage* (Exeter: David & Charles, 1990).
2   George Monbiot, *Regenesis: Feeding the World Without Devouring the Planet* (London: Allen Lane, 2023).
3   Thom Dallimore, 'Movers and Shakers: adaptation and survival in soil organisms', *WoodWise* Spring 2016 (Woodland Trust, 2016).
4   Suzanne Simard, *Finding the Mother Tree* (London: Allen Lane, 2021).
5   Peter Wohlleben, *The Hidden Life of Trees: What they Feel, How They Communicate* (Vancouver: Greystone Books, 2016).
6   Merlin Sheldrake, *Entangled Life: How Fungi Make Our Worlds, Change Our Minds and Shape Our Futures* (London: Bodley Head, 2020).
7   Brooke Action for Working Horses and Donkeys, 'How Working Horses Shaped Britain', https://www.thebrooke.org/news/how-working-horses-shaped-britain#:~:text=Farm%20work,ploughing%2C%20tilling%20and%20hauling%20manure.
8   DEFRA, *Keepers of Time: Ancient and Native Woodland and Trees Policy in England* (DEFRA, 2022).
9   Wendell Berry, 'A Native Hill', *The World-Ending Fire* (London: Allen Lane, 2017).

10 Guy Standing, *Plunder of the Commons: A Manifesto for Sharing Public Wealth* (London: Pelican, 2019).

Chapter Eleven: Wilding the Woods

1 Isabella Tree and Charlie Burrell, *The Book of Wilding: A Practical Guide to Rewilding, Big and Small* (London: Bloomsbury, 2023).
2 https://www.woodlandtrust.org.uk/trees-woods-and-wildlife/animals/mammals/wild-boar
3 https://www.rewildingbritain.org.uk/why-rewild/reintroductions-key-species/key-species/eurasian-beaver
4 https://www.wildlifetrusts.org/saving-species/beavers#:~:text=The%20Scottish%20Beaver%20Trial%20is,of%20the%20Landfill%20Communities%20Fund.
5 Isabella Tree and Charlie Burrell, *The Book of Wilding: A Practical Guide to Rewilding, Big and Small* (London: Bloomsbury, 2023).
6 Guy Shrubsole, *The Lost Rainforests of Britain* (London: William Collins, 2022).

Chapter Twelve: The Long Spring

1 Planet Ark, *Wood: Housing, Health, Humanity* (Planet Ark, 2015).
2 Yoshifumi Miyazaki, *Walking in the Woods: Go Back to Nature With the Japanese Way of* Shinrin-Yoku (London: Aster, 2021).
3 Aldo Leopold, *A Sand County Almanac* (reissued, Oxford: Oxford University Press, 1968).
4 Robin Wall Kimmerer, *Braiding Sweetgrass: Indigenous Wisdom, Scientific Knowledge and the Teachings of Plants* (London: Penguin, 2020).
5 Nicola Chester, *On Gallows Down: Place, Protest and Belonging* (London: Chelsea Green, 2022).
6 Nick Hayes and Jo Moses, eds, *Wild Service: Why Nature Needs You* (London: Bloomsbury, 2024).
7 Wendell Berry, 'A Native Hill', *The World-Ending Fire* (London: Allen Lane, 2017).
8 Forestry Commission, *Key Performance Indicators Report 2022–23* (Forestry Commission, 2023).
9 Forestry Commission, *Provisional estimates of the ownership type and property type of woodlands in Britain* (2020). https://cdn.forestresearch.gov.uk/2022/02/fr_nfi_woodland_ownership_report_srf3wpm.pdf

10 Guy Shrubsole, *Who Owns England?* (London: William Collins, 2019).
11 Patrick Barkham, 'Britain ranks bottom in Europe for nature connectedness', *Guardian*, 23 June 2022.
12 Tori Tsui, *It's Not Just You: How to Navigate Eco-Anxiety and the Climate Crisis* (London: Gallery UK, 2023).
13 Robin Walter, *Living With Trees* (Beaminster: Little Toller Press, 2020).
14 Oli Rodker, *The Cutting Edge: Twelve Examples of Ecological Forestry in the UK* (Landworkers' Alliance, 2023).
15 Ian Tubby and Adrian Jowitt, 'Working together to meet government targets for tree planting in England', Natural England and Forestry Commission blog, 8 April 2024 https://naturalengland.blog.gov.uk/2024/04/08/working-together-to-meet-government-targets-for-tree-planting-in-england/
16 James Rebanks, *English Pastoral: An Inheritance* (London: Allen Lane, 2020).
17 Henry Dimbleby, *National Food Strategy: The Plan* (National Food Strategy, 2021).
18 Climate Change Committee, *Land Use: Policies for a Net Zero UK* (Climate Change Committee, 2020).
19 Patrick Joyce, *Remembering Peasants: A Personal History of a Vanished World* (London: Allen Lane, 2024).
20 Joe Moran, *First You Write a Sentence: The Elements of Reading, Writing... And Life* (London: Viking, 2018).
21 Sue Clifford, 'Up the Forest!', in Adrian Cooper, ed., *Arboreal* (Beaminster: Little Toller Press, 2016).

Epilogue

1 Jim Dodge, 'Introduction', Gary Snyder, *The Gary Snyder Reader* (Berkeley: Counterpoint, 1999).

Acknowledgements

ONE OF THE colleagues I spoke to while researching *Ancient* reminded me, in the friendliest possible way, that I'd better get it right because I'm doing it for all the other woodlanders who'll never get the opportunity. I'm very conscious that I owe a debt of gratitude to countless people who've shared their insights and expertise with me during more than twenty years in the woods. A full list would be a book in itself, so for all the brilliant woodlanders I've met who aren't listed below: this is for you.

As a ranger, I worked with and learned from great colleagues at Cheshire Countryside (particularly Darren Evans and the late Gareth Seel); Dunham Massey; the City of London City Commons; South Lakes; the Peak District; and through being involved with the Ancient Tree Forum – thank you all. Leigh Cawley trained and inspired an entire generation of National Trust ranger apprentices at Reaseheath College. The influence of Jamie Simpson should be obvious from the text, and anyone looking for an expert woodlander in the southeast should contact him. Big up bruv.

More recently, Ludo Cinelli at the Eve White Literary Agency took a gamble on me and found *Ancient* a home. Along with Steven Evans, Ludo also contributed enormously to refining the structure and message of the book, as well as to the ongoing process of getting it right. Thank you.

I've been fortunate to receive the attentions of three brilliant editors at Profile books in Cecily Gayford, Jenny Dean and Grace

ACKNOWLEDGEMENTS

Pengelly, and each of them has improved the book immeasurably, along with copy editor Jacqui Lewis. Thanks also to Jon Petre, Georgina Difford and everyone else involved with the book's production and promotion.

John Deakin, Head of Trees and Woodland at the National Trust, supported me in taking a sabbatical to write *Ancient* and read various drafts. He is also simply a great boss, who I'm proud to work with. Rosie Hails, Ian Goode, Tom Hill, Matt Stanway and the rest of the trees and woodland team have all played a part in helping this book take shape in one way or another – I feel very lucky to work with all of you and the rest of our NT colleagues. A special shout out to the amazing community of rangers. Also at the Trust, David Elliot, Kev Davies, Dan Cameron, Simon Rogers, Tom Wood, Iain Carter, Felicity Roos, Rachel Oakley, Sue Cornwell and Jamie Lund shared their deep knowledge of their sites and subjects with me, read sections and provided helpful comments.

One of the many joys of my day job is working with experts from other organisations, too. Specifically in relation to the book, at the Woodland Trust my conversations with Saul Herbert have contributed enormously to the way I articulate the need for management of ancient woods, along with the work of his former colleagues Alastair Hotchkiss and Christine Reid, and current ancient woodland leads Adrian Southern and Dean Kirkland.

At Natural England, Joe Alsop and Dan Abrahams taught me all about ash dieback and the White Peak woods, while Nigel Pilling, Marion Bryant, Neil Ford, Paul Mortimer and Emma Dear clarified various details on top of the expertise they routinely share. Andrew Weatherall is a staunch advocate for ancient woodland at the RSPB and was also, in a previous life, my MSc tutor. Thank you to everyone involved with the England Woodland Biodiversity Group, the Wood Pasture and Parkland Network, and the UK Rainforest Network.

Ancient woodland pioneers George Peterken, formerly of

the Nature Conservancy Council, and Charles Watkins, of the University of Nottingham, were more generous than I could ever have expected in conversation and in reviewing some of the text. If it isn't already obvious, their books, along with the work of Oliver Rackham and the other woodland books in the 'Further Reading' section, form the core texts from which I've soaked up much of the general material in *Ancient* over the years.

Donald and Sheila Pigott were incredibly kind and supportive in our few meetings before Donald's death in 2022. Baz Gutteridge, Shaun Waddell, Dave Watson, Thom Dallimore, Tom Coxhead, Tom Kemp, Helen Pitman, Matt Hayes, Andy Poore, Daniel Brown, Rosie Rendell, Tom Burns, Elena Pearce and Brian Williamson all helped bottom out the detail and reviewed relevant passages and chapters. Clive Mayhew was kind enough to let me use his original research into the role of ancient woodland in the two world wars – thank you.

As well as getting me hooked on small-leaved lime, Hugh Milner helped pursue some of the mysteries of the species as well as reading relevant chapters. Jonathan West sadly died in 2018 but his children, Bede and Fritha – both brilliant environmentalists in their own right – answered my questions and read the chapter featuring him.

Conversations with Guy Shrubsole helped me work out some of my thinking about belonging and land rights. Guy also provided thoughtful advice on the publishing process, along with Patrick Barkham and Lee Schofield.

Space forbids me from listing the many non-woody friends who've supported my work on this book in one way or another, but much love to the Marple crew, 21CC, Time Team (RIP) and our amazing network in Sheffield. Thank you to my parents for instilling in me a love of both the outdoors and of reading, and for everything else along the way. And finally, thank you Megan, for your love, patience and support, and, of course, Wren and William. I hope to leave a woodier world for you.

Index

Act for the Preservation of Woods (1543) 200
agriculture 9, 17, 20, 26, 31, 78, 81, 97, 101, 106, 108, 120, 128, 161, 171–2, 176, 177, 180, 195, 219, 221, 233, 234, 240, 249, 250, 252, 253, 256, 266, 268; agricultural depressions 141, 169; Common Agricultural Policy 279, 280; future of 276–84; industrialisation of/intensive 12, 40–41, 82, 138–41, 146, 157, 244; regenerative 41, 224, 254, 282; wood pasture and 73–6
air pollution 172, 185
alder 56, 66, 129, 165, 191, 251
Alsop, Joe 160, 163–4, 167, 169–71, 176–7, 180–81, 190–91
Ancient Semi-Natural Woodland (ASNW) 14, 18, 20
Ancient Tree Forum (ATF) 96, 240, 251
ancient woodland: active management 4, 134, 140, 153–8, 203, 205–10, 217, 235–7, 254, 259, 263, 274; age of, determining 13, 15–16, 111; ancient trees 86–9, 94, 97, 102, 110, 113, 220; climate change and *see* climate change; conservation, early attempts at 10–11, 17–18, 21; coppicing and *see* coppicing; deer and *see* deer; definition of 10–17, 76–7; ecological condition of 4; felling *see* felling trees; indicators of 5, 15–17, 113, 114, 166, 168–9, 171; 'Locust Years'/destruction and abandonment of (1945–1970s) 3, 9–11, 17, 19, 128, 134, 145–7, 170–71, 177, 233, 262; Plantations on Ancient Woodland Sites *see* Plantations on Ancient Woodland Sites; protected from clearance 3–4, 10–11, 17–19, 19*n*, 50, 75, 80, 81, 112, 235, 237, 240, 248; records refreshed and renewed (2019)/(2025) 19–20; rewilding and *see* rewilding; soil *see* soil; threats to 4, 18–19, 159–91 *see also individual threat name*; wildwood *see* wildwood; wood pasture *see* wood pasture
Ancient Woodland Inventory (AWI) 17–19, 76, 229, 253
Anglo-Saxons 24, 73, 130–31, 184, 217
archaeology 32, 41, 106–32, 144, 153, 166, 210, 233, 235, 255–6
Arnold, James: *The Shell Book of Country Crafts* 201–2
ash ix, 9, 25, 33, 42, 62–6, 75–6, 121, 127, 129, 198, 199, 208, 230, 255; dieback (*Hymenoscyphus fraxineus*) 5, 160–71, 180–91, 267
Ashridge Estate 175, 234
Ashtead Common 5, 55, 69–72, 75–8, 80, 83–7, 92–7, 102–6, 108, 110, 138–9, 149, 150, 153, 164–5, 192, 194, 210, 212, 243, 262, 272, 281, 283
Attenborough, David 85
auroch 98, 102, 130, 246

Barnby, Vernon Willey, 2nd Baron 80
bats 86, 133–8, 152–8, 210; Barbastelle 137, 155; Bechstein's 137; noctule 136; pipistrelle 134–7, 152–3, 156
beanpole 34, 35, 39–40, 54, 146, 261
beaver 98, 100, 241, 245, 247, 251–3, 290
bees 59, 60, 86, 102, 246
beetles 16, 85–7, 99, 102, 153, 185, 187, 211, 219, 245
belonging, sense of 119, 262–4, 269–70, 274, 275, 284, 290
Berry, Wendell: *A Native Hill* viii, 120, 121, 234, 237, 270, 284
Białowieża, Poland 239–45, 254, 256, 258–9
billhook 30–34, 40, 89, 266, 282
binders 34, 39, 40, 42, 115, 260
bioabundance 60
biodiversity 60, 103, 149, 177–8, 236, 242, 244, 252, 259, 273
biomass 177, 208, 209, 218, 274
birch 25, 42, 44, 45, 58, 62, 65–6, 69, 83, 104, 129, 143, 244, 261

303

birds, woodland 59, 60, 83, 95–7, 105, 133, 138–40, 153, 155–6, 174, 177, 178, 184–6, 189, 211, 230, 234, 241–3, 251, 255, 256, 264, 276, 280, 290. *See also individual species name*
bison 103, 241, 245–8, 258, 270
Black & White Building, East London 208
Black Down 202, 203
blackbird 95, 133, 189
blackcap 42, 59, 139, 155, 190
blast furnace 126–7
Blean Woods National Nature Reserve 151, 246–7, 258, 262, 270
bluebell 8, 10, 15, 20, 42, 55, 62, 67, 169, 189, 232, 249
blue tit 155, 185
boar, wild 98, 104, 241, 247–51, 258, 290
bobbins 14, 127, 141
bolling 72, 77, 84, 86–7, 89, 105, 111
bramble 4, 30, 42, 55, 60, 83, 92, 104, 129, 148, 150, 152, 165, 175, 216, 219, 226, 229, 232, 238, 248, 257, 287, 290
British Association for Shooting and Conservation 177
British Trust for Ornithology 59
broadleaved trees 12, 13, 19, 96, 206, 230, 235, 238, 257, 271
Bronze Age 2, 21, 31, 32, 41, 73, 101, 103, 106, 114, 123, 195, 200, 216, 226, 283
Brown, Capability 81
Brown, Daniel 216–17, 225–9, 238
bull pits 102
Burrell, Charlie: *The Book of Wilding* 97, 244
butterflies 2, 42, 54, 59, 60, 94, 97, 133, 138, 147–52, 155, 161, 186, 213–15; chequered skipper 147; fritillary 60, 147–9, 151–2; purple emperor 97, 101, 105, 243; red admiral 59, 150
Butterfly Conservation 59, 151

California, US 21, 28, 45–53, 190, 289
Canada 45, 47, 141
canopy 62, 69, 83, 98, 100, 101, 108–10, 117, 121, 124, 149–51, 159–62, 165, 170–71, 174, 192, 194, 203, 211–12, 214, 233, 239–43, 246, 255–7, 285, 286; defined 59–60; species 87, 162, 187, 217
carbon 12, 37–9, 172, 189, 207, 208, 221–2, 236, 238, 263, 273, 276, 281, 282
Carter, Iain 231
carving 260–61, 275
caterpillars 60, 87, 147, 148, 150, 185–6, 192, 223
cathedrals 2, 33, 194, 196, 197, 203
cattle 3, 70, 77, 92–4, 99, 104, 231, 243, 244, 246, 247, 255–6, 281
chainsaw ix, 4, 31, 42, 46, 49, 58, 63, 64, 94, 152, 160, 197, 232, 234, 256–7
charcoal 3, 14, 17, 23, 33, 37–9, 56, 68, 107, 118, 124–6, 140, 142–3, 166–7, 195, 200, 230, 285, 287, 288
Charter of the Forest (1217) 75
cherry tree 25, 62, 65, 178
Chester, Nicola: *On Gallows Down* 129, 269
chicken of the woods fungus 84–6, 210
choker-setting 218, 225
City of London Corporation 27, 30, 31, 38, 42, 54, 79, 80, 104
Clifford, Sue 284
climate change 4, 16, 100, 169, 183–6, 189, 217, 222, 258, 267, 271–3, 282, 284, 288–9
climax community 98
coke 39, 126, 140
commodification/marketisation of nature 236
Common Ground 284
commons/commoning 12, 27, 36, 69–91, 93–5, 101, 103–105, 231, 236, 237, 272, 273
Commons Preservation Society 79–80
community groups 261–2, 264, 267, 270–74
conifer 9, 13, 17, 18, 19, 56, 141, 143, 146, 147, 170–71, 217, 229–33, 246, 271
Coniston Water 108, 116, 126–7, 129, 285
coppice 2–5, 12, 13, 17, 22–43; billhook and 30–34, 40, 89, 266, 282; birds and *see* birds, woodland; butterfly declines and 147–53; charcoal and *see* charcoal; commercialisation of 24; commons and 78, 82; coppice-crafts 23, 41, 201–2, 232, 260–62, 269, 274; coppice-worker communities 260–64, 269–71, 274; coupe *see* coupe; defined 2, 23; earliest evidence of in Britain 32; fall in area of woodland under coppice management 3, 17, 57, 72, 140–44; firewood and *see* firewood; hurdles and *see* hurdles; language of 24; 'overstood' 31, 56, 84, 86, 107–9, 110, 113, 119, 129, 144, 206–7, 217, 225; poles 22, 24, 26, 31–5, 75, 77, 84, 86–90, 108, 109, 111, 119, 122, 128, 142, 217–19, 221, 225–8, 277–8; products 2–3, 4, 13, 23, 25, 27, 33, 35, 37–40, 43, 57, 62, 78, 127, 140, 141, 147, 235, 236, 261–3, 275, 278; process 22–3, 30–31, 33–8, 39–43; restored coppice woods 23–5, 31, 43, 54–62, 147–54, 209–10, 214–15, 217, 231, 234, 237, 254, 260–64, 269–71, 274, 275, 284; standards *see* standards; stools *see* stools; tree evolution and 61, 68; wattle and daub and 41; wildwood and 61–2; World Wars and 140–44. *See also individual wood name*
coupe (coppice worksite) 24–6, 31, 36–7, 39, 41–2, 44, 54–5, 57–60, 62, 64, 67–8, 104, 111, 131, 138, 150–52, 193, 203, 218, 225, 227, 238, 263–4
Croft Castle 209, 231
cruck blades 199–200

INDEX

Dallimore, Thom 220
Dallington Forest 56, 262
Davies, Kev 243–4
decaying wood 49, 51–2, 61, 65–6, 71–2, 85–7, 99, 102, 114, 153, 186, 194, 210, 212, 219–22, 240–41, 247, 251, 256–7, 287, 288
declines, wildlife/species 3, 21, 59, 60, 92, 95, 105, 134, 138–40, 148–9, 151, 153, 155, 174 241, 247, 254, 263, 271, 289
deer 4, 12, 15, 29, 42, 80, 81, 98, 99, 112, 113, 123, 129, 147, 160, 173–7, 179–80, 185, 189, 206, 231, 243–4, 247, 250, 263–4
Dodge, Jim 289
Dodgson Wood 6, 107–108, 115–16, 119–21, 124–30, 151, 161–2, 168, 173–5, 184, 195, 212, 253, 255–7, 265, 286, 288, 290
Douglas fir 145, 231
Dovedale Wood 5, 159–91, 253, 255, 266, 267, 286
drought 182–5, 222
Dunham Massey 29–30, 53, 82, 96, 267

Eastern Moors Project 257
Ebworth 155
ecosystem services 12, 189, 236, 237, 272
Eddie's Wood 7–9, 11, 13–17, 20–21, 27, 269
electric fencing 41, 104
Elliott, David 202–3
elm x, 117, 170, 181, 187, 200; Dutch elm disease 187
Ely Cathedral 197
enclosure x, 72, 74–5, 77–9, 82, 83, 91, 105, 106, 236, 251, 272
English Heritage 127
Ennos, Roland: *The Wood Age* 35
Epping Forest 27, 79
estovers 75, 80, 273
Etherow Country Park 20
Evans, Darren 28
exclosure 128
extinction 50, 61, 98, 101, 148, 149, 157, 245, 246, 249, 289

faggots 36, 201
Fairlie, Simon 225
fallen trees 52, 85, 114, 117–19, 153, 257, 285–9
felling trees xi, 2, 16, 18, 36, 39, 44–68, 94, 119, 132, 144, 145, 147, 151, 201, 204, 211, 215, 217, 227, 233, 252; clear-felling 54, 154–5; selective felling 154–5, 210, 274, 275, 284
fertilisation 12, 83, 224
fertilisers 39, 40–41, 76, 161, 171, 222, 242, 277
field maple 170, 181, 190, 191, 198
firewood 3, 14, 23, 33–9, 67, 75, 77, 80, 86, 167, 205–7, 209, 273–5, 278, 281
First World War (1914 18) 140–43, 145, 245

Flag Fen 32, 226
Ford, Neil 253
Forest of Dean 14, 82–3, 127, 201, 249, 250
Forest Research 147
forestry: agroforestry 282; commercial or 'scientific' 9, 13, 141, 142–5, 147, 257; continuous cover forestry (CCF) 154–6, 274; forestry harvester 229, 233, 234; forestry plantations 3, 17, 78, 82, 229, 233, 282; male profession 268; skills shortage 270
Forestry Commission 4, 18, 20, 108, 142, 143, 145, 146, 147, 230, 232, 262; *Utilisation of Hazel Coppice, The* (bulletin, 1956) 146
forests 21, 28, 45, 47, 80; deforestation 209; forest bathing 264, 265; forest browsing 73; forest law 80–81; temperate rainforest 56, 116–17, 129, 143, 151, 164–5, 173, 195, 200, 212, 231, 253, 255–6; term 80
Forge Wood 56–60, 67, 111, 153, 179, 204, 207, 234, 262, 263–4, 268, 270
foxglove 58, 59, 60, 68
Francis-Baker, Tiffany: *The Bridleway* 96, 238
Fuller, Rob 59
fungi 1, 12, 16–17, 36, 49, 66, 84–5, 102, 172, 183, 185, 189, 195, 211, 219–24, 229, 240–42, 244
Furness Abbey 126–7

garden warbler 59, 155
ghylls 56, 116, 121, 124, 126, 130, 168, 286, 288
glades 61, 70, 104, 151, 155, 157, 177, 243, 245, 247, 282
Gloucestershire Wildlife Trust 250
Gow, Derek 251
grazing 3, 5, 12–14, 40–41, 61, 67, 70, 72–7, 81, 83–4, 93, 94, 98–100, 128, 130–31, 147, 162, 168–70, 172, 174, 220, 223, 231, 242–4, 255–6, 279
Green, Ted 96
grey squirrel 206, 250
Gutteridge, Baz 42, 45, 53–4, 59, 62, 64, 65, 67, 153, 218

habitat: evolution, megafauna and 61, 68, 100–101; habitat piles 36, 58; interactions between 95; kaleidoscope of 100; open 93, 95, 99, 103
Hainault Forest 79
hammer pond 56–7
Haskell, David George 105
Hawkcombe Woods 151, 152
hawthorn 25, 69, 83, 93, 94, 102, 104, 170, 232, 257, 266
hay/haymaking 83, 138, 172, 276–7, 278; tree hay 75, 96, 281–2

305

Hayes, Matt 246
hazel 3, 22-7, 30-32, 34, 36, 39-43, 54-9, 62, 101, 104, 108, 113, 117, 123, 124, 126, 131, 133, 140, 146, 170, 174, 175, 193-5, 199, 217, 251, 253, 262, 264, 278
headage payments 278
hedge/hedge-laying 30, 34, 42, 82, 95, 138, 260, 266, 278, 279, 281
Herdwick sheep 127, 130, 277, 278
Himalayan balsam 179, 180, 256
holly 8, 25, 104, 113, 117
hornbeam 56, 58, 162, 198, 239
horse-logging 216-17, 225-9, 234, 238
housing 1-2, 3, 27, 33, 39, 106, 108, 193, 195, 196, 202
HS2 18
Humboldt State University 47, 50
hurdles 40-42, 123, 126, 262, 274, 278

Ice Age x, 11, 100, 101, 162, 246, 288
indicators, ancient woodland 5, 15-17, 113, 114, 166, 168-9, 171
indigenous people 49, 121
Industrial Revolution 3, 33, 57, 72, 105, 133, 140, 173
insects 42, 60, 67, 104, 136, 137, 138, 149, 155, 156, 230, 290
Iron Age 30, 76, 106, 167, 226
iron industry 14, 30, 39, 56, 73, 126-7, 140, 151, 179, 195, 196, 200, 201

Jamie, Kathleen 120
Joyce, Patrick: *Remembering Peasants* 283

Kent Wildlife Trust and the Wildwood Trust 246
keystone species 4, 98, 247, 258
Kimmerer, Robin Wall: *Braiding Sweetgrass* 117, 267
King Alfred's cakes fungus 160, 187, 189
Kirby, Keith 6, 16
Knepp Estate 96-8, 100, 240, 242-4, 248, 254, 258, 276

landowners 18, 50, 73-5, 78-9, 81, 141, 156, 174, 177, 198, 236-7, 271-2
Landworkers' Alliance 270-71, 274
larch 129, 142, 145, 208, 231
Lathkill Dale 160, 170, 177
Law, Ben 108, 202
Leeds Coppice Workers 270-71
Leopold, Aldo 4, 267
lichens 16, 56, 66, 69, 116, 117, 173, 189, 212, 253-6
lime x, xi, 33, 41, 61, 65, 113-25, 127, 130-32, 144, 162, 166-8, 171, 181, 190, 191, 225, 238, 255, 265, 285-90; age of 113-15, 288; collapse and resurgence 285-9; large-leaved 110, 132, 162, 190;

Pigott and *see* Pigott, Professor Donald; reproductive strategies 16, 113-15; small-leaved x, 5, 16, 101, 107, 108, 109-10, 114, 116, 121, 131, 166, 184, 217, 240, 285, 302; species native to Britain 110; survival 130-32; symbolism 132; *Tilia* genus and 115, 121-2; uses/lime-bast cord 121-4; woodbanks and 111-14
Linnard, William 74
livestock 3, 12, 14, 32, 34, 40-41, 62, 70, 72, 75-7, 81, 83, 92-4, 99, 104, 112, 130, 140, 146, 147, 157, 168, 172, 185, 190, 231, 243, 244, 246, 247, 250, 255-6, 277, 281, 283
Lodgepole 47-51, 53, 190
London 1, 5, 26-7, 30, 54-6, 79-80, 92, 94, 103, 107, 111, 115, 120, 130, 149, 205, 208, 213, 268
Lover's Leap 159, 163-4, 182, 187
Lund, Jamie 126
lynx 173, 176, 241, 245, 250, 258

Magic Map 20, 76, 127
Magna Carta (1215) 75, 77
Makers Barn 261-2, 263
Maramureș, Romania 156-8
Marple 7-8, 20
Marren, Peter 6, 99, 110
Mary Rose 200
Mayhew, Clive 142, 145, 198
megafauna 61, 100-101, 103
mental health 119-20, 264-7
mezereon 166
mills 8-9, 13-14, 127, 172, 198, 205, 206, 231
Milner, Hugh 108
minimum intervention 153
Monbiot, George: *Regenesis* 220
monocultures 9, 13, 73, 131, 141, 142, 145, 168, 170, 189, 214, 242
Moran, Joe 284
Mosquito fighter-bomber 143
mountain currant 166, 187

National Food Strategy 281
National Nature Reserve 151, 155, 177
National Trust 5, 6, 28, 29, 53, 79, 107, 126, 128, 131, 155, 171, 174, 175, 182, 202, 209, 222, 230-31, 234, 237, 243, 255, 270, 276, 281
natural capital 236
Natural England 19, 76, 119, 160, 168, 177, 253
Nature Conservancy Council 10, 14, 17; Ancient Woodland Inventory *see* Ancient Woodland Inventory (AWI)
navies 33, 83, 141, 194, 200
Neolithic period 31, 41
Nibthwaite 126-7
Niemann, Derek 18
nightingale 60, 95-7, 101, 105, 139, 147, 243
nitrogen pollution 172-3, 177, 185

INDEX

non-native species 13, 18, 57, 140, 142, 143, 145, 178, 179, 204, 230, 232
Norman Conquest 23, 32, 74, 80
North West Mull Community Woodland Company 273-4
Notre-Dame Cathedral, Paris 196-7, 205

oak ix-x, 5, 8, 9, 20, 26, 29, 54-5, 58, 65, 93, 94, 97-8, 101-2, 109, 116-18, 142-5, 151-3, 162, 164, 170, 174-5, 184-5, 191, 192-215, 230-40, 243, 244, 253, 262, 283, 285, 287, 288; acute oak decline 185; age of 86; climbing 69-70, 84-5, 87-91, 194, 197, 199, 210-14; coppicing and 22, 24-7, 29, 33; felling 54, 55, 57, 58, 65-6, 202-6; pegs/'trenails' 195-6; pollard 5, 84, 104, 272; pruning 87-91, 94, 104, 192, 193, 194, 200; standards 24-5, 27, 33, 57, 124-5, 135-6, 144, 193, 198-9, 202, 210, 235, 275; timber, use as 127-9, 194-201; uses of 201-3; wood pasture and 69-72, 75, 76
Oostvaardersplassen 99
Open Spaces Act 27, 80
Ötzi (Bronze Age mummified man) 123-4

pannage 75, 273
parks 12, 49, 50, 51, 53-5, 80-83, 110, 112, 113, 172, 174, 258
Pearce, Elena 102
Peel Island 125-6
Peterken, George 6, 10, 11, 12, 15, 17-18, 21, 23, 59, 74, 130, 146, 147, 198, 257
pheasant 177-8, 271
phenolic compounds 264-5
Pigott, Professor Donald 6, 115-16, 119, 128, 132, 171, 255
pine marten 50, 250-51
pioneer species 162-3
pit props 141-3
Pitman, Helen 246
plantation 3, 9, 12-13, 17, 19, 56-8, 78, 80-83, 96, 129, 142-6, 175, 205-6, 217, 257, 274, 282
Plantations on Ancient Woodland Sites (PAWS) 19, 129, 229-35, 246, 271
planting trees 11, 13, 18, 145, 146, 154, 180, 232, 276-80; in ancient woodland 170-71
pollarding 3, 5, 56, 58, 72, 74-7, 79, 81, 83-4, 86-8, 90-94, 96, 104, 107, 110-13, 115, 161, 164-5, 192, 194, 207, 212, 232, 272, 277
pollen 99, 101, 121, 187, 264
Poore, Andy 155
Prince's Trust 267

Quinlan, Rina 100, 248

Rackham, Oliver 2, 6, 9-13, 17, 21, 27, 35, 54, 61, 70, 74, 80, 86, 98, 111, 114, 128, 134, 145, 147, 171, 175, 196, 233, 257, 273; *Trees and Woodland in the British Landscape* 10
Ransome, Arthur 125-6
Read, Sir David 223
Rebanks, James 280, 281, 283
redwood 21, 28, 45-53, 190
refugee habitat 96, 139
reintroduction, species 4, 151, 176, 241, 245-6, 250-52, 258
Rendell, Rosie 262
Repton, Humphry 81, 82
rewilding 93, 98, 100, 102, 103, 239-59, 273
rhododendron 57, 178-80, 204, 267
rides (linear clearings along access routes) 61, 66-7, 144, 148, 149, 150, 151, 152, 177, 282
Right to Roam movement 269
roading 36, 143
robin 95, 248, 266
Robinson, Tony 168
Roman Empire 30, 32, 73, 76, 195-6, 226
Roos, Felicity 222, 282
Rotherham, Ian 106, 169, 233
Rushmore Estate 155, 274

sallow 97, 104
Samson, Lisa: *Epitaph for the Ash* 189
San Francisco, US 28, 46, 47, 49, 50
saproxylic invertebrates 85, 102
Second World War (1939-45) 3, 9, 13, 17, 21, 27, 54, 93, 128, 134, 143, 144, 167, 217, 227, 229, 277
secondary woodland 12, 19
shadow woods 169
Sheldrake, Merlin: *Entangled Life* 223
Sheringham Park 214
ship-building 2, 33, 39, 83, 123, 140, 141, 194, 200, 201
Shrubsole, Guy 116; *The Lost Rainforests of Britain* 256
Simard, Suzanne 223
Simpson, Jamie 11, 55-9, 67, 87-9, 93, 96, 111, 126, 134, 135, 152, 153, 178-9, 204-6, 217, 228, 232, 234, 240, 243, 263-4, 268, 270
Snyder, Gary 46, 187-8, 289; *Rip-rap* 48
soil 216-38; carbon in 221, 222, 236, 238; fungi/Wood Wide Web and 36, 222-4, 229, 244; government grant payments and 237; hand-cutting trees and 231, 234-5; horse-logging and 216-17, 225-9, 238; humus layer 219, 221, 222; importance of ancient 222-5; invertebrate species and 219-22; living experiment, as 224-5; marketisation of natural systems and 236; natural capital and 236; Plantations on Ancient Woodland Sites restoration and 229-38

spalting 66
spoon-carving 260-61, 270, 274, 275
Spring Park 5, 23-8, 30, 31, 33, 36-7, 39, 42-3, 53-7, 59, 62, 66-8, 79, 107-12, 115-17, 121-2, 124, 130, 131, 133, 135, 137, 144-5, 149-53, 158, 164-5, 184, 193-4, 196, 203, 209, 210, 212, 214-19, 225, 234-5, 238, 262, 273, 275, 283
stakes 34, 35, 39, 40, 42, 54, 231, 260, 262, 278, 281
standards 57-9, 142, 164, 195-6, 200-202, 212, 271; oak 24-5, 27, 33, 57, 124-5, 135-6, 144, 193, 198-9, 202, 210, 235, 275
stools 24, 30, 31, 41, 42, 54, 57, 61, 107, 108, 109, 110, 113, 130, 144, 145, 146, 174, 217-18, 229, 230, 238
storms 99, 100, 122, 183, 184, 257-8
Stott Park 127
succession 98, 183, 215
sweet chestnut 26-7, 29, 54, 56, 57, 65, 108, 109, 131, 140, 144, 145, 184, 193-4, 202, 209, 216-18, 221, 225, 228, 229, 234, 238, 246, 260
Sweet Track 32

tanbark 127, 140-41
tanneries 124, 127, 142
Tansley, Sir Arthur 167
Tarras Valley Nature Reserve 273
Teign Valley 231, 233
Threehalfpenny Wood 109, 111
timber: British imports 140-42, 207-8; continuous cover forestry and 154, 155; defined 24; engineered 208; harvesting in ancient woodland 209-10, 234, 235, 275; knee timber 199-201, 208, 275; lime 121; oak 33, 128, 193, 194-208, 215; origins 31; plantations 9, 12, 18, 145-6, 154, 282; redwood 49; traditional uses 201-2; World Wars and 140-44
tree time 188, 243
Tree, Isabella: *The Book of Wilding* 97, 244
treen 23, 33, 261
turbary 75

UK Butterfly Monitoring Scheme 148, 149, 151
underwood 24, 33, 56, 83, 127, 128, 142, 162, 240
University of California, Berkeley 28, 46, 47, 189

vegetative reproduction 15, 16, 57, 162
Vera, Frans 98-100, 244
Victory, HMS 200

Waddell, Shaun 104, 105
Walter, Robin: *Living With Trees* 62, 274
Warde, Paul 35
Warren, Martin 59
Watkins, Charles 6, 11, 18, 140

Watson, Dave 122-3, 232
wattle and daub 3, 41, 195
West Blean Woods 246
West, Jonathan 108
Westonbirt Arboretum 40
Whiddon Deer Park 112
White Peak 5, 156, 160-62, 164, 167, 171, 176, 180-82, 188-90, 265-6
Wild Ennerdale 242, 244
wildflowers 5, 7, 12, 13, 20, 54, 66, 94, 101, 107, 147, 160, 161, 165, 168, 170-72, 190, 231, 233, 243, 280, 282, 290
Wildlife Trust 237, 246, 250, 271
wildwood 2, 5, 11, 14, 61, 68, 73, 93, 98-102, 105, 107, 114, 121, 130, 137, 155, 158, 168, 171, 184, 190, 194-5, 241, 246, 247
Williamson, Brian 40, 123, 232
willow 32, 69, 83, 92, 93, 95, 97, 104, 195, 246, 257, 261, 265
willowherb 58, 60, 290
willow warbler 42-3, 59, 92, 95, 101, 138, 139, 155, 283
Windsor Great Park 82, 85
Witherslack 151-2
withies or weavers 40, 41, 195, 262
wolf x, 130, 173, 241, 248, 250, 258
women 143-4, 268, 271; Women's Timber Corps 143-4
wood anemone 15, 20, 42, 55, 67, 113, 120, 165, 168, 238, 264
woodbank 17, 58, 59, 106, 107, 111, 112, 125, 130, 131, 166, 217, 225, 230, 234
wood colliers 37, 125, 126, 198
wood culture, resurgence of 5, 157-8, 242, 260-63, 290
Wood, Jacqui 123
wood pasture 3-5, 20, 53, 56, 61, 70-83, 86, 103-4, 110, 112, 131, 175, 194, 231-2, 255, 276-7, 280, 283; clearance and simplification of 138; commons and 12, 72-80; defined 12, 70; habitat interaction in 95; humans as wood pasture species 105; livestock and 281; origins/history of 70, 72-83, 105, 265, 272; rewilding and 93, 96-9, 244-5, 281; wildlife and 70-71, 80-81, 85-6, 93-9, 137, 139, 157, 173, 281
wood warbler 129, 139, 185, 186, 255
woodchip 218, 251, 274
woodland, creating new 276-7
Woodland Trust 18, 230-31, 237; *State of the UK's Woods and Trees* report 19
woodmanship 9, 33, 45, 62, 126, 201; woodwright 201-4
Working Woodlands Cornwall 271
Workman, John 155

Yeo, Sophie: *Nature's Ghosts* 100
yew 159, 166, 167, 190
Yorkshire Wildlife Trust 271